OUR TOOL-MAKING SOCIETY

IRENE TAVISS

is a Lecturer in the Department of Sociology, Harvard University. From 1966 to 1972 she was Research Associate to the Harvard University Program on Technology and Society. She is the editor of *The Computer Impact,* co-editor (with Everett Mendelsohn and Judith Swazey) of *Human Aspects of Biomedical Innovation,* and the author of numerous articles.

OUR TOOL-MAKING SOCIETY

Irene Taviss

PRENTICE-HALL, INC. · ENGLEWOOD CLIFFS, NEW JERSEY

Library of Congress Cataloging in Publication Data

Taviss, Irene.
Our tool-making society.

(A Spectrum Book)
Includes bibliographical references.
1. Technology and civilization. I. Title.
HM221.T345 301.24'3 72-4053
ISBN 0-13-644484-9
ISBN 0-13-644476-8 (pbk.)

10 9 8 7 6 5 4 3 2 1

Prentice-Hall International, Inc. (*London*)
Prentice-Hall of Australia Pty. Ltd. (*Sydney*)
Prentice-Hall of Canada, Ltd. (*Toronto*)
Prentice-Hall of India Private Limited (*New Delhi*)
Prentice-Hall of Japan, Inc. (*Tokyo*)

TO JULIE

Contents

Preface

In their original form, the essays that follow appeared as introductions to the Harvard University Program on Technology and Society *Research Review* series (distributed by the Harvard University Press). They were intended as state-of-the-art summaries. For purposes of this volume the essays have been expanded and revised, but they retain much of their original tone as overviews of the issues and controversies in the field.

In the course of writing them I have been aided and influenced by the many people who have worked with the Program, as well as by the scholars whose works I review. I owe a special debt of gratitude to Emmanuel Mesthene, who has served as counsellor and as "editor"—in the fullest and best sense of that word.

OUR TOOL-MAKING SOCIETY

INTRODUCTION

PERSPECTIVES ON THE TECHNOLOGY–SOCIETY RELATIONSHIP

> Using stones and branches as he found them, and then devising new combinations of these natural materials, [man] gradually made other environment changes. Descending a river on a tree trunk, or a log hollowed out by his new tool, fire, or maybe a raft, was as natural as harnessing the wind to propel his canoe or ship. If this proposition is sound, there is nothing unnatural in the progression that led to a steel ship propelled by twin screws driven by a steam turbine. . . . As the man-made sectors of human environment grow in importance, technology, rather than being merely an activity of man, partakes more and more of the nature of man. The urge to alter the conditions of life and to improve them in directions of his own choosing is clearly a basic instinct.[1]

In a time that is all too filled with romantic notions about the "natural state of man" and the "dehumanization" wrought by technology, it is useful to remind ourselves that man is a tool-making and tool-using animal. Man is also a social animal who bands together with other men to accomplish common purposes and who passes on his traditions and learning to his offspring. His values—conceptions of the desirable—and his knowledge become embodied in his culture. His methods of cooperation become embodied in his social structure—the organization of society into institutions, strata, and roles. His technology—the tools and organized knowledge that he uses to achieve his purposes —is an integral part of his culture and social structure. In accordance with his values and social structure, he develops technologies, which in turn have consequences for his society and values.

Because the second-order consequences of technological change may be unanticipated, discrepancies may arise between particular technologies and other elements of the social system. Aspects of the social structure and values of a society may also be discrepant with each other at some points in time. From this vantage point, the study of technology and society is an examination of the processes of mutual adjustment among technologies, social structures, and values.

Often a technological change that is implemented for one purpose or goal has secondary and unintended consequences that interfere with other goals. Thus, the technologies that allowed for increased agricultural productivity and helped produce affluence had the undesirable effect of displacing large numbers of unskilled workers who migrated to urban areas and helped to generate the poverty and congestion of the inner city. New biomedical technologies that served the goals of health and well-being by providing better medical care generated an increased specialization and division of labor within medicine that upset the older pattern of many general practitioners and a few specialists. As a result, the psychological comfort, general medical advice, and knowledge of the "whole patient" that were provided by the general practitioner were interfered with.

American society is particularly beset with social problems

resulting from technological change, because it is a society that places a high value on the rapid development of new technologies. "Probably more than any other society, we revere technological innovation, . . . [and] the assumption that the new will be better than the old goes very deep in our culture." [2] It is part of the American mythology that "if a man can make a better mousetrap, the world will make a beaten path to his door."

Until relatively recently, critics of technology in American society were few and generally not taken seriously. The obvious benefits of technology in the form of labor-saving machines and a higher standard of living seemed clearly to outweigh whatever spiritual costs the critics decried. But in the aftermath of World War II, a disaffection with technology began to set in, largely owing to the development of the atomic bomb. Concern began to be expressed about the social responsibility of scientists and technologists and the socially harmful consequences of their work. The virtues of an expanding science and technology were no longer taken for granted. Some ten to fifteen years later, the consequences of technological change for employment became a focus of much attention and concern. "Technological unemployment" became a rallying cry for labor leaders, as well as many others who viewed rapid technological change with suspicion. The "automation hysteria" of the 1950's and early 1960's was stilled as evidence became available that technological change would not render large segments of the population unemployed and as automation progressed less rapidly than had been predicted. More recently, the environmental dangers produced by large-scale technology in combination with a high population density have brought a deeper disillusionment. Ecological hazards and environmental degradation have become the new rallying cry for those who fear the deleterious consequences of technology. And a kind of "alienation hysteria" has replaced the earlier "automation hysteria." Critics proclaim that a society characterized by rapid technological change is destructive of human values and renders man alienated and impotent in the face of a dominant and autonomous technology. The new wave of disenchantment with technology has brought with it an attack on the very roots of technological progress—the scientific, rational, empirical-experimental stance toward the universe.

The attack on both technology and rationality comes at a curious time in history. In the days before the Industrial Revolution, technological development tended to be relatively haphazard. By the twentieth century, the decision to develop a new technology became deliberate and highly self-conscious. Yet ironically, the more deliberate and self-conscious man has become in developing technologies and attempting to control their effects, the more technology has come to appear as dominating and autonomous.

Why is this so? To begin with, the slower development and smaller scale of technology in the past made its consequences less disturbing and its effects less massive. Hence, technology did not assume the aura of an all-encompassing and autonomous social force. Second, the very attempt to control technology brings consequences of its own. In an advanced technological society, the awareness of the possibility of altering both technologies and social structures has made the continued existence of social problems seem less necessary than in the past. At the same time, the complexities of a densely populated heterogeneous modern society have made the solutions to social problems more difficult than ever. The inability to solve technology-related social problems helps give an appearance of autonomy to technology.

The dominance of technology and its harmful social consequences have become a popular theme of recent social criticism. Often, technology is linked with or used as a virtual synonym for the "cult of efficiency," the "military–industrial complex," or the "Brave New World." Because of the problems and uncertainties surrounding the uses and effects of modern technologies, technology lends itself easily to such symbolic use and acquires an aura either of "the devil" or of "awe-and-wonder." To be sure, insofar as technology is not separable from social structures and values, the connections between it and such social phenomena as the "cult of efficiency" are real. Of course, the connections between technology and affluence or between technology and increased longevity are no less real, and supporters of technology argue that many current problems could be alleviated through better applications of technology.

TECHNOLOGY AND THE SOCIAL SYSTEM

The extremes in the current debate about the merits of technology tend to ignore or underemphasize the intrinsic connections between technology and the society of which it is a part. From the perspectives of the antitechnologists, technology is either an extrinsic force that should be expunged or a dominant part of the social system whose evil influences can be abolished only by overthrowing the entire social system. Technology is thus often portrayed as the culprit in situations where institutional inadequacies are the root cause of the problem. The extreme technological optimists are equally guilty of failure to comprehend the interaction between technology and society. They fail to recognize that technology, as a part of a complex social system, cannot by itself bring a cure for all social ills.

The nature of the relationship between technology and society has long been a subject of theorizing and debate. The chief issue under contention has been whether technology is the principal determinant of social structure and values or vice versa. In the Marxian schema, the "mode of production" determines all social life and is the motor force of history. But the mode of production is itself highly dependent on the forces of production—that is, on technology. Thus, Marx argued: "In acquiring new productive forces, men change their mode of production; and in changing their mode of production, in changing the way of earning their living, they change all their social relations. The handmill gives you society with the feudal lord; the steam-mill, society with the industrial capitalist." [3] Technological change produces social change because new technologies at a certain stage in their development come into conflict with the existing organization of production and exchange. Hence, Marx argued, just as the feudal society was transformed into the bourgeois society, the technology of the factory system will ultimately be incompatible with the social structure of the bourgeois society. The transformation into socialism will occur through the resolution of the "contradiction between socialized production and capitalistic appropriation," between "socialized organization in the individual factory and social anarchy in production as a whole." [4]

Rejecting Marx's determinism, Max Weber argued that values play a crucial role in social change. According to Weber, capitalism was not an automatic outgrowth of feudalism in the way that Marx thought it was. Rather, capitalism required the prior emergence of a specific set of values, those entailed in Puritanism. Economic factors alone could not account for the development of the rationalistic capitalistic system. The "ability and disposition of men to adopt certain types of practical rational conduct" was crucial, since "when these types have been obstructed by spiritual obstacles," rational economic conduct fails to emerge.[5]

There is a voluminous literature on the controversy between Marx and Weber that cannot be reviewed here. Currently, although adherents to both sides remain, a "middle ground" seems to be gaining strength in the form of what William James called a "soft determinism." The argument is that while technology is a principal determinant of changes in social organization and values, the precise effects of a given technology will depend upon the preexisting structures and values. As Robert Heilbroner has succinctly summarized it, "We cannot say whether the society of the computer will give us the latter-day capitalist or the commissar, but it seems beyond question that it will give us the technician and the bureaucrat." [6]

The determinism of Marxian theory seems inadequate today, in part because of the heightened awareness of the possibilities of shaping and developing technology for social purposes. As Heilbroner and others have pointed out, technological progress is itself a social activity, present to a high degree in some societies and relatively absent in others. Furthermore, the course of technological advance is responsive to social policies and values.

"Whether technology advances in the area of war, the arts, agriculture, or industry depends in part on the rewards, inducements, and incentives offered by society." [7]

While values and social structure block or promote the development and adoption of various technologies, they also help to mediate the effects of technology. For example, although technology exerts a major influence on the nature of the work that men do, the social structure and values of an industry or a given firm help to determine the effects of technology on the worker and the work process. The same technology may be used in different ways, so that extreme division of labor may prevail in one place, while in another the managerial philosophy will dictate a program of "job enlargement."

Yet technological change also confronts the social structure and value system with new issues and demands. Technological change has, of course, been the root cause for the transformation of an agrarian society into a densely populated, complex industrial society. The governance of such a society, its basic institutions and values, and the psychology of its individual members are all affected by the changes technology has brought and continues to bring. The following chapters seek to illuminate these effects and to explore the interrelationships among technologies, values, and social structures in the context of contemporary social issues.

Technology is a facet of most social issues in modern societies, although its role varies from minor to major. In matters of work and leisure, of health and education, of transportation and housing, technology is both a partial cause of problems and potentially a partial solution. In matters of social and economic justice, definitions of the valuable and the good life, and the maintenance of an effective and democratic government, the role of technology is more indirect. But the existence of advanced technologies helps to set the social context in which such issues are framed and dealt with.

Also helping to set the context for most social issues is the well-known phenomenon of "rising expectations." There is a heightened awareness of unmet social needs and a heightened expectation that ameliorative social action will be taken. Yet such action is extremely difficult—in some cases owing to ignorance about how to effect meaningful social change, in other cases because of value conflicts, and in still others because it would require a rearrangement of power and authority and of incentive and reward structures. Hence, the many false starts toward alleviating social problems often serve only to increase frustration.

RATIONALITY REVISITED

Frustration with the inadequacies of social policy and the continued existence of social problems has characteristically led to proposals for reform. Recently, however, the very basis of social decision making has come under suspicion. What is being challenged is not simply "the power structure" or "vested interests," but the method of decision making itself. While a more complex society would appear to call for a more sophisticated application of rationality, rationality has become a target.

Critics charge that scientific rationality is subverting higher human values. Some erstwhile supporters of rationality have begun to wonder about its limitations and deficiencies. Defenders of rationality, meanwhile, maintain that true rationality has never been applied to social decision making and therefore, like Christianity, has never been tested.

Clearly, rationality did not appear for the first time in the twentieth century, the Industrial Revolution of the nineteenth century, or the Enlightenment of the eighteenth century. Yet its scope has been vastly extended, in much the same way that technology in the twentieth century has become qualitatively as well as quantitatively different from previous technology. The more radical among contemporary critics of technological society argue that the ultimate culprit is not technology itself, but the mentality that goes with it: the precise, calculating, quantitative rationality that obliterates both subjectivity and human values. They argue that no sphere of existence is free from the demands of objectivity and scientific reasoning, that no other way of "seeing" has validity in our culture. A culture whose dominant tone is rooted in objectivity and rationality places so much emphasis on the cognitive aspects of human existence that it ignores or stifles the more affective and emotional domains. This leads to the crippling of human personality and the demise of meaning and value. Hence, the "counterculture" proclaims the need to overthrow the scientific world view. In its stead, the counterculture proposes a system based on subjective insights and visionary experiences.[8]

While the somewhat mystical withdrawal advocated by the counterculture is essentially alien to the Western tradition, the attack on rationality is not. Such classic modern Western theorists as Weber and Mannheim expressed concerns similar to those of contemporary critics, although their critiques of rationality were more subtle.

Weber distinguished between "formal" and "substantive" rationality, Mannheim between "functional" and "substantial" rationality. Both saw the former trait as peculiarly characteristic of modern societies. Formal rationality—the whipping boy of the counterculture—is the methodical calculation of the means to achieve a given end. Substantive rationality involves an assessment of "the content of the particular given ends to which it is oriented."[9]

In Mannheim's terms, the problem of modern rationality is that while "increasing industrialization . . . implies functional rationality, i.e., the organization of the activity of the members of society with reference to objective ends, . . . it does not to the same extent promote substantial rationality, i.e., the capacity to act intelligently in a given situation on the basis of one's own insight into the interrelations of events." There is, rather, a tendency for functional rationality to have a "paralysing effect . . . on the capacity for rational judgment."[10] Mannheim argued that society should be organized in such a way as to facilitate substantial rationality. He believed that a democratically planned society would help to achieve this by clarifying both the relationships among groups within the society and the relationship of individual to collective action.

For Weber, the essence of modernity was to be found in the progressive "demystification" and "rationalization" of the world. Rationalization is the displacement of an "unthinking acceptance" of custom and tradition by a "deliberate adaptation to situations in terms of self-interest." This process, he contended, "can proceed in a variety of . . . directions; positively in that of a conscious rationalization of ultimate values; or negatively, at the expense not only of custom, but of emotional values; and, finally, in favour of a morally sceptical type of rationality, at the expense of any belief in absolute values." [11] Weber was concerned about finding institutional means of promoting positive forms of rationalization.

The counterculture rejects the very possibility of a positive form of rationalization. Their rejection of rationality or "linear" thinking represents a total rejection of the existing social system and a call for a new society based on love and subjective truths.

While the counterculture does not have a large number of adherents, elements of it do seem to strike a responsive chord in many individuals. The most important reason for what resonance the counterculture does generate is probably the sheer size and complexity of modern social and technological systems. Individuals facing such large and highly interdependent systems often find it difficult to understand them or to feel any sense of potency vis-à-vis their operation. The political system, for example, is often perceived as being either chaotic, buck-passing, and fragmented, or hypercontrolled, conspiratorial, and repressive. In either case, control does not seem possible. Because the rationale that appears to underlie the system is incomprehensible, it often takes on a magical and malevolent aspect, and the fully operational rationality allegedly at its base becomes a principal symbolic target.

The current attack on rationality thus takes three forms: an attack on rationality as a symbol for what is remote and incomprehensible; a concern about the pitfalls of functional or formal rationality and its tendency to overpower substantial or substantive rationality; and a reaction against an upset in the balance between the affective and cognitive aspects of life in favor of the latter.

Some modern critics attempt to combine elements of both the last two. Erich Fromm, for example, maintains that modern man suffers from a "low-grade chronic schizophrenia" characterized by the "growing split of cerebral-intellectual function from affective-emotional experience," which arises in the attempt to cultivate scientific, "emotion-free thinking." [12] As an example of such "emotion-free thinking," he cites the work of Herman Kahn on thermonuclear warfare. "The question is discussed: how many millions of dead Americans are 'acceptable' if we use as a criterion the ability to rebuild the economic machine after nuclear war in a reasonably short time so that it is as good as or better than before. Figures for GNP and population increase or decrease are the basic categories in this kind of thinking, while the question of the human results of nuclear war in terms of suffering, pain, brutalization, etc., is left aside." [13]

Such attacks constitute only one part of the current challenge to rationality. There is a second part that is rooted in the sense that modern society is so overwhelmingly complex that rationality is inadequate to the task of decision making and control. This idea has been expressed in extreme form in a recent work of science fiction. *The Tale of the Big Computer* postulates a "Sociological Complexity Theorem," which states that "the problem of organizing society is so highly complex as to be insoluble by the human brain, or even by many brains working in collaboration." [14] The basis of this dictum is that "only after a thorough study of social conditions has been made can one take steps to organize society in a rational way. But it takes a long time to grasp all the elements of a complex social structure. If a community . . . remains static, one has leisure to study its essential characteristics, decide what measures are necessary, and adopt them in due order, testing one's way toward a satisfactory system. . . . [B]ut one cannot apply the methods used in a static society . . . to a society which has been rendered progressive through technical development; for during the time taken to study the community, work out necessary measures and then implement them, science will have made new discoveries and technology new inventions, so that at the moment when innovations begin to take effect they will already be outdated." [15]

In less extreme form, the complexity theorem is frequently enunciated by social analysts, candid or frustrated government officials, and reformers of varying ideological stripe. These people share a disillusionment with the promise of rationality. As one analyst has noted, "No established institution in our society now perceives itself as adequate to the challenges that face it." [16] In any process of decision making, the number of variables that may be relevant or even crucial makes "the resulting matrix . . . too complex to display and evaluate. Efforts at pruning and simplification of variables for purposes of experiment tend to have the usual effects of the Procrustean bed—one ends up talking about a situation that may be manageable but has little intrinsic interest. . . . The conclusion is . . . that in the process there must be certain non-rational steps. It is only on a non-rational basis that one can make the leap from the virtually infinite combinations of possible variables to some finite set." [17]

As portrayed in science fiction, the solution to the Sociological Complexity Theorem lies in computer technology. When the computers assume responsibility for all decision making and for the performance of all social functions, life runs smoothly and everybody is happy. Outside the realm of science fiction, attempts to cope with the complexities of modern society often take the form of proposals for new techniques of decision making that rely heavily on computers. Such proposals rest on the assumption that more sophisticated applications of rationality to issues of public policy will go a long way toward solving our social problems. Thus, for example, Jay Forrester proclaims that sophisticated computer modeling of such complex systems as cities will give us the wisdom to make the right governing decisions.[18] The required decisions, he asserts, are "counter-intuitive" and hence could not be arrived at by any other means. Aside from the pitfalls of Forrester's model itself, his assumption that the method is of central importance gives ammunition to the counterculture.

The "new utopians"[19] who believe that society can be transformed simply through the use of their techniques are ignoring social and political realities. While computers and the new techniques of decision making may facilitate increased rationalization of government decision making, there are strong countervailing forces and considerations. Like any technology, computers do not exist in a vacuum. The preexisting political structures and traditions help to determine how they will be used. Considerations of power and political gain, as well as organizational inertia, often block the effective use of new techniques of decision making. Because the new utopians fail to take into account such existing social structures and values, as well as the limitations of the techniques themselves, many of them are as naive in their way as members of the counterculture are in theirs. Both groups tend to offer unrealistic panaceas as an escape from complexity.

They tend, moreover, to feed on each other. The more loudly the new utopians proclaim the wonders of their techniques, the more the counterculture reacts to the horrors of "rationality." The two are extremes, of course; more moderate positions exist along the spectrum between them. But their ideological fervor often has the effect of polarizing the community.

Problems associated with the uses and consequences of rationality affect not only our governing processes and methods of dealing with social problems, but also our values and life-styles. The consequences of technological change have included an increase in both the complexity of society and the degree and scope of rationality. As a consequence of the rising expectations that accompany a technology-based affluence, government is faced with a greater challenge of providing for social needs and solving social problems. At the same time, the techniques of accomplishing these goals have become more suspect as a richer and more educated population exerts increased demands for citizen participation and power.

Rationalization of the political process is further complicated by value conflicts and uncertainties. The goals to be attained are not self-evident or undisputed. Furthermore, a highly complex technological society is one in which the very basis of values becomes less firmly grounded and shared. The affluent modern man questions established values and searches for his identity. Yet his expectations and demands for meaningful work, better health care, and more livable cities grow. The chapters that follow attempt to shed light on these complicated interrelationships.

□ □ □

1. R. J. Forbes, *The Conquest of Nature: Technology and Its Consequences* (New York: Praeger Publishers, Inc., 1968), pp. 57–58.
2. Kenneth Keniston, *The Uncommitted: Alienated Youth in American Society* (New York: Harcourt Brace Jovanovich, Inc., 1965), pp. 213–14.
3. Karl Marx, *The Poverty of Philosophy* (New York: International Publishers, 1963), p. 109.
4. Friedrich Engels, "Socialism: Utopian and Scientific," in Lewis S. Feuer, ed., *Marx & Engels: Basic Writings on Politics and Philosophy* (New York: Anchor Books, 1959), pp. 95 and 110.
5. Max Weber, *The Protestant Ethic and the Spirit of Capitalism*, trans. Talcott Parsons (New York: Charles Scribner's Sons, 1958), pp. 26–27.
6. Robert L. Heilbroner, "Do Machines Make History?" *Technology and Culture*, 8 (July 1967), 342.
7. *Ibid.*, p. 343.
8. See Theodore Roszak, *The Making of a Counter Culture: Reflections on the Technocratic Society and Its Youthful Opposition* (New York: Doubleday & Company, Inc., 1969).
9. Max Weber, *The Theory of Social and Economic Organization*, trans. A. M. Henderson and Talcott Parsons, ed. Talcott Parsons (New York: Oxford University Press, Inc., 1947), p. 185.
10. Karl Mannheim, *Man and Society in an Age of Reconstruc-*

tion (New York: Harcourt Brace Jovanovich, Inc., 1940), p. 58.

11. Weber, *op. cit.*, p. 123.
12. Erich Fromm, *The Revolution of Hope* (New York: Bantam Books, Inc., 1968), pp. 41–42.
13. *Ibid.*, pp. 42–43.
14. Olof Johanneson, *The Tale of the Big Computer* (New York: Coward-McCann, Inc., 1968), pp. 19–20.
15. *Ibid.*, p. 18.
16. Donald A. Schon, *Beyond the Stable State* (New York: Random House, Inc., 1971), p. 17.
17. *Ibid.*, p. 215.
18. Jay W. Forrester, *Urban Dynamics* (Cambridge, Mass.: The M.I.T. Press, 1969).
19. Robert Boguslaw, *The New Utopians: A Study of System Design and Social Change* (Englewood Cliffs, N.J.: Prentice-Hall, Inc., 1965).

TECHNOLOGY and the POLITY

The most basic issue confronting modern democratic states is how to maintain a government that is both effective and democratic. Technological change has exacerbated this long-standing problem by changing the social context in which effectiveness and democracy are defined and measured. It has done so principally by increasing social complexity and interdependence. Technological change also impinges more directly on the polity through the new techniques of decision making that it has made possible.

THE INDIRECT RELATIONSHIPS: IMPLICATIONS FOR EFFECTIVENESS AND DEMOCRACY

A complex and highly interdependent society seems to require a greater degree of planning and coordination than did the simpler societies of the past and to depend more on knowledge and expertise. If the government intervenes more in the direction of society, the criteria of effectiveness are broadened and conceptions of democracy cannot remain the same.

Technological change has been indirectly responsible for these developments in three ways. First of all, by bringing about a more elaborate division of labor, it has produced a greater interdependence among men, a larger number of interacting organizations, and a heightened need for the coordination of these separate activities and organizations. Second, the use of modern technologies makes the actions of individuals and organizations affect the well-being of others to a greater degree than in the past —whether they pollute the atmosphere through industrial wastes, drive automobiles, fly airplanes, or even use their stereo equipment. Hence, there is greater need for the regulation and control of private activities. Third, and more indirectly, technology-based affluence has increased social interdependence and the need for public decision making by producing a shift in the composition of demand in favor of collective goods and services that cannot be purchased privately and are not provided effectively by the market. The provision of such goods as clean air and water can be made only by the public sector, through the political process. The desire for more public goods, the need to control the effects of modern technologies, and the problems of coordinating a complex division of labor have all given rise to a need for greater social control, for more governmental intervention. As a consequence, the traditional assumptions on which American democracy has rested are being challenged.

The Challenge to American Traditions

The free reign of the market, the "hidden hand" of laissez-faire equilibrium, and the rationality of individual decision makers have long been the guidelines for American government. Technological and social change have rendered all three of these inadequate. The increased importance of public goods and the need for controls on the uses of technology have limited the scope of the market mechanism. The "hidden hand" is made inoperable when business, government, and research institutions produce such potent technologies as nuclear weapons, computers, artificial organs, and supersonic aircraft, and when giant organizations can control the economic fate of large segments of the population. Finally, the rationality of individual citizen decision makers is insufficient to deal with the complexities that such developments have brought. For "the growth of numbers of people, amounts of knowledge, and speed of change in technology work against the individual being in a position to exercise free, reasonably well-informed, rational, individual choice concerning much of his destiny. . . . As technology grows, markets expand, and societies grow in size, the individual's share of the knowable decreases drastically [and] the limitations of the individual become more marked relative to society as a whole." [1]

The traditional American image of the public arena is that of a marketplace in which competing groups struggle to maintain or secure power, while the government acts as a kind of broker among them; public policies emerge from this bargaining process rather than being deliberately formulated by government. In an age of large-scale technology and giant organizations, this image is inadequate. As the power of private organizations has grown, the government too must enlarge its power if the liberty and participation of the citizenry is to be protected.

> The classic notion [of American government] was that rights inhered in individuals. But the chief realization of the past thirty years is that not the *individual* but *collectivities*—corporations, labor unions, farm organizations, pressure groups—have become the units of social action, and that individual rights in many instances derive from group rights, and in others have become fused with them. Other than the thin veil of the "public consensus," we have few guide lines, let alone a principle of distributive justice to regulate or check the arbitrary power of many of these collectivities. . . . [There is thus a] lack of any institutional means for creating and maintaining

> the necessary public services. On the municipal level, the complicated political swapping among hundreds of dispersed polities within a unified economic region, each seeking its own bargains in water supply, sewage disposal, roads, parks, recreations areas, crime regulation, transit, and so on, makes a mockery of the *ad-hoc* process.[2]

In the absence of deliberate social planning and control, private interests reign and the public as a whole has little say in decisions that affect them. Thus, "private economic interests" continue to make such decisions as "what industries will eliminate labor and what human functions will be replaced by the computer, which is to say they will decide what the future social and economic class system will look like. . . . There is simply no basic acceptance of the legitimacy of such decisions being made by conscious action of the whole community rather than by the incremental decisions of individuals. . . ."[3] The absence of such legitimacy, the fact that social planning and a strong government are inimical to American political traditions and values, constitutes a major stumbling block to change.

The provision of more effective government through greater coordination and planning and the increased reliance on technical knowledge inherent therein seem to pose problems for democracy. The government is often deemed to be ineffective when it fails to solve social problems, yet the fear of undue government power and of the excessive authority of technical experts often makes it difficult for government to assume the responsibility for planning and control that might allow it to prevent or ameliorate such problems. What appears to be required, then, is a new conception of democracy to match the new conception of effectiveness. "The nature of government for a multi-billion person world . . . is neither quantitatively nor qualitatively the same as that required for an isolated New England village. What freedoms do we intend to preserve? Perhaps it would be more accurate to ask: What new concepts of freedom do we intend to attach the old names to?"[4]

Because the operations of American government lie somewhere on the continuum between laissez-faire and planning, contemporary assessments of the power of government run the gamut between those who fear too little control and those who fear too much. The range of views may be illustrated by the following summary: While Peter Drucker proclaims the "sickness of government" and its inability to function adequately in its expanded role,[5] Michael Harrington decries the "accidental century" that has resulted from the failure of government to assume such increased responsibility,[6] and Henry Kariel similarly attacks the insufficiencies of the "limited, impotent state" that have given rise to the reign of private power and political apathy;[7] while Herman Kahn and Anthony Wiener fear the advent of a totalitarian state because social complexity and new technologies will lead to an increase in government power and the temptation to give power to "a new Caesar," [8] Victor Ferkiss fears "chaos rather than hypercontrol" as public purposes give way to "the feudalistic struggle of competing special interests";[9] while Daniel Greenberg proclaims the "myth of the scientific elite," [10] Jean Meynaud sees political power shifting from elected representatives to technocrats;[11] while Robert Boguslaw foresees a new "computerized bureaucracy" that will have the power to impose its values on governmental decisions,[12] James Schlesinger sees the political process as rendering the new techniques of government ineffective or inoperable.[13]

There seems to be some agreement, however, that the trend is toward planning and the assumption of greater governmental power and responsibility. As noted, this is alien to our traditions and values; for while Europeans were asserting the sovereignty of government after the Industrial Revolution, Americans were rejecting it. "To the extent that sovereignty was accepted in America it was held to be lodged in 'the people.' Popular sovereignty, however, is as nebulous a concept as divine sovereignty. The voice of the people is as readily identified as is the voice of God. It is thus a latent, passive, and ultimate authority, not a positive one." [14]

If the notion of planning for the collective welfare were to attain legitimacy, problems of implementation would, of course, remain. Planning would not eliminate the political bargaining process or the conflicts of interests and values that characterize a large and complex society. Indeed, planning is likely to increase such conflicts. By setting up a "specific locus of decision," it would offer "a visible point at which pressures can be applied" and thus make conflicts more open.[15] Nor will popular demands for a say in the decision-making process be diminished. To the contrary, "the expansion of governmental intervention in the economic and social life of the nation increases the stakes of participation: the government does more, and, therefore, more is to be gained by having a voice over what it does." [16]

In addition to the value problems, planning also entails difficulties stemming from the inadequacies of our knowledge. These are of two principal sorts, one of which may be called "social" and the other "technical." Our social knowledge is unable to provide us with adequate guidelines for how to achieve the communal coordination that a system of democratic planning requires. Our technical knowledge is often inadequate in determining the best means to accomplish those ends that society agrees to foster. The dangers that are seen as latent in a movement toward greater government planning and power are intimately related to these difficulties.

If planning leads to a commitment to long-range programs that are based on faulty knowledge, the consequences could be grave. Decisions and policies whose tenure might run through several successive administrations would be less reversible than in a non-planned system, and hence error would be more serious in its consequences. Insofar as the techniques for arriving at and implementing a social consensus about goals are inadequate, there is the danger that both ends and means might be imposed on society under the guise of expertise. If a commitment to planning is made while there is no effective method for choosing goals, there is some danger that the planners might make arbitrary decisions and sell these to the people on the basis of their superior knowledge. If the public maintains exaggerated notions of the skills and powers of technical experts, either a self-fulfilling prophecy could emerge or a revolt against the experts could appear that would result in "throwing out the baby with the bathwater." That is to say, an exaggeration of the claims to authority based on expertise could bring about either an overhasty adoption of plans and programs that have been insufficiently examined, or an across-the-board rejection of technical expertise that could lead to the destruction of socially beneficial programs.

The Public and the Private Sectors

Given the current limitations on the power of government to intervene in social matters, the development and uses of technology pose problems for government effectiveness. Because most new technologies are developed by private industry, technological progress has tended to be most rapid in the production of goods and services that can be sold on the market. Since the market mechanism is not an adequate regulator of public goods, production of such goods has tended to lag behind the production of consumer goods.

A "pure" public good is one that is consumed jointly and on an all-or-none basis. Clean air, for example, benefits all indiscriminately. Thus, the firm that incurs the cost of producing such goods cannot "sell" them in the usual way. "The market therefore provides no effective indication of the optimal amount of such public commodities from the point of view of society as a whole." [17] It redounds to the government, therefore, to insure an adequate provision of such goods. The government also bears heavy responsibility for those "mixed" goods, such as health care and education, that are provided by both public and private sources and are marketed in accordance with public specifications.

It is difficult to find mechanisms that will work as well in the production and distribution of public goods as the private sector works in producing and distributing private or consumer goods.

With both "the discipline of the profit-and-loss statement" and "the clear feedback and the incentive of calculable rewards" lacking,[18] mechanisms have not yet been developed that are comparable to those used by the corporate sector to search out and develop new goods and services, and to distribute them.

Attempts have been made to tap the skills and resources of the corporate sector for public purposes. In addition to awarding contracts and grants for research and development to private corporations and nonprofit organizations, the government has attempted to create markets for public goods and thus provide incentives to the private sector. But such attempts have not met with much success, with the notable exception of military and space goods, where the market is clearly guaranteed and there is no problem of distribution. In areas such as health and education, the institutional complexities, government regulations, and vested interests have operated to block the process.

A more radical approach has been suggested by Peter Drucker. After noting the inertia of government bureaucracies, he recommends a policy of "reprivatization"—of returning the responsibility for society's goals and tasks to private institutions. According to this scheme, government would formulate choices and set policy, but would turn over the actual execution and performance of the tasks (the "doing") to nongovernmental institutions. He argues that business is particularly appropriate for this because "it is predominantly an organ of innovation; of all social institutions, it is the only one created for the express purpose of making and managing change." [19]

This proposal does not depreciate the importance of government or assign it a minimal role. There is no assumption about a "withering away of the state." To the contrary, Drucker argues that "reprivatization" would require a strong government to manage and coordinate. For those who bemoan the ineffectiveness of current government, it is an attractive proposal. However, since governmental regulation would clearly be required to ensure that the activities of private institutions do in fact serve the public good, might not the government thwart some of the efficiency and innovative capacity that the private sector can now maintain precisely because it does not have to function in the public interest?

Numerous other proposals have been propounded to ensure representation of the public interest in the decisions and activities of the corporate sector, the most celebrated of which are those of Ralph Nader. All these proposals involve a change in the restrictions or incentives that currently operate in the private sector.

New restrictions or incentives are also necessary to take account of the "external" costs of technology—those negative consequences that do not redound to the producer or user. The pollution of our air, water, and sound waves has not been adequately controlled because our society has operated on a modified laissez-faire system in which the freedom of individual or corporate decision making has been restricted only minimally, and therefore such social costs have tended to be neglected.

The "external" costs of technology could be "internalized" if they were to be charged directly to the producer—that is, if the polluters or those who benefit from their activities were made to pay the costs. "Technology would seem much less autonomous if we could internalize all its costs and charge them directly to the producer. They would then be paid for either by the ultimate consumer (in the form of higher prices, in the case of consumer goods, or of some sort of fee, as in a toll-road) or by society as a whole when necessary (in one form or another of public subsidy), but only following a deliberate and explicit public decision to do so." [20] A second, and complementary, method for controlling pollution would be to control it at its source (for example, by developing electrically powered automobiles) or to develop new technologies capable of counteracting the pollution (such as special filters).

Internalization of costs might also be achieved through the more gradual workings of the judicial process. An extension of the existing legal doctrines of nuisance, strict products liability, and liability for harm from "abnormally dangerous activities" could serve this end.[21] In extending these doctrines, the intent is to penalize industries for failing to take into account the harmful effects of their technologies. The private sector would thus be forced to assume social responsibility in its initial decision making with respect to new technologies. The penalties should not be so great, however, as to stifle research and the development of beneficial new technologies.

THE DIRECT RELATIONSHIPS: IMPLICATIONS FOR EFFECTIVENESS AND DEMOCRACY

The use of new techniques of decision making to carry out the expanded functions of government has been a source of concern. Current fears that the reliance of government on technical expertise is turning the United States into a "meritocracy" or a "technocracy" are not without historical roots, however. There has long been a strain in the American value system stemming from the inconsistencies between the goals of "equality" and "achievement." If "achievement" is rewarded too highly, "equality" may suffer, and vice versa. The balance between the two has shifted back and forth during the course of American history.[22] Current fears about "meritocracy" may thus reflect a concern over a new and undue shift in the balance toward the "achievement" side in an advanced industrial society.

There are some analysts who do not recognize any dangers in "meritocracy" and do not see it as incompatible with equalitarian democracy. Brzezinski, for example, argues that "equal opportunity" can coexist with "special opportunity for the singularly talented few." He says that we are becoming a "meritocratic democracy," which combines a continued respect for the popular will with an increasingly important role for specialists in the key decision-making institutions.[23] But others see "meritocracy" and "democracy" as inherently contradictory, because current social trends "encourage a political and cultural gap between the upper and lower ends of American society. . . . [T]hese can now be characterized as those who manage and those who are managed by advanced technological systems. Nevertheless, with rising skill and educational levels, the "managed" are becoming more interested in and capable of self-management.[24] In the face of this contradiction, can one expect some shift back toward the "equality" end of the spectrum?

Technocracy

Perhaps a prior question is the extent to which the American political system has become or is becoming "technocratic." With the exception of Meynaud, most analysts of the decision-making process do not believe that we are tending toward a technocracy. As Nieburg has put it, despite a "formalization of the scientific role at the highest levels, the trend has been away from the acceptance of scientific authority, toward the assimilation of this new tribe of experts into the ambiguities of traditional politics. In essence the scientists . . . are neither sloe-eyed vestal virgins of abused innocence nor pristine high priests of a new technocracy. . . . Influence over public policy is nebulous and difficult to measure, but it is clear that, whatever its constituents, scientists individually or as groups have not possessed inordinate amounts."[25]

But the concern that undue power may accrue to those experts who control information by their ability to program the computers and build the decision-making models is not to be dismissed lightly, even though the evidence to date indicates that experts are not "taking over." The possibility remains that information overload coupled with attention scarcity and the complexity of issues could lead to an abdication of political authority to technical experts. A more likely scenario, however, is one in which experts and politicians collaborate to achieve their power ends.

Numerous commentators have noted that the "neutrality" of experts is not to be taken for granted. As information centers for decision-making purposes are established, information scientists may not be able to avoid becoming protagonists in political competitions. Take the following as an example: Suppose that an elected official promotes "a costly public-works project as a way of stabilizing his political position. If the analyst should fail to include those so-called 'noneconomic variables' in calculating expected returns from the investment, he misses the main point. If he does account for them, he aligns himself with the interests of one partisan group against others." [26] Insurance against such partisan tendencies requires the deliberate airing of "alternative action and value hypotheses." [27]

Both those whose principal concern is the improvement of government effectiveness and those whose principal concern is the preservation of democracy share the hope that alternative possibilities will be properly exposed. The former seek mechanisms whereby some form of competition, pluralism, and/or adversary procedures can be built into the uses of information systems and associated analytic techniques. The latter seek mechanisms to assure that the information upon which decisions are based will be open to the public, that dissenting sources and types of information will be exposed, and that the underlying assumptions will be made explicit and available for public scrutiny.

An example of the importance of adversary procedures has been provided by Alain Enthoven, a former Assistant Secretary of Defense for Systems Analysis. In 1966, the Secretary of the Army and the Joint Chiefs of Staff had recommended to the Secretary of Defense that an antiballistic missile be deployed to defend American cities against Soviet attack. (The more recently proposed ABM systems are for defense of our intercontinental ballistic missile sites.) Both the army and the Systems Analysis Office produced elaborate cost-effectiveness studies. As might be expected, the army's estimates were much more favorable toward the proposed missiles. By request of the Secretary of Defense, the two groups and their heads met to examine and argue out their differences. After careful recalculations and reexamination of assumptions and methods, it was discovered that there was one decisive difference between the two analyses: The army assumed that the Russians would continue to deploy their projected offensive forces, while the Systems Analysis Office assumed that they would react by deploying more missiles or other devices to overcome the opposing defenses. Using the latter assumption, calculations showed that the antiballistic missiles would not be effective, and the Secretary of Defense argued against their deployment.[28]

It is the deliberate concealment of such partisan interests that is of concern to those who fear that an alliance of experts and politicians will undermine democratic government. Because of the difficulty of determining true costs and benefits, many programs and plans have always been uncertain. In the face of such uncertainty, the politician could often decide in such a manner as to preserve and extend his power, and "the politician and professional could cover up these . . . facts of life quite well most of the time." What is worrisome is that "in the future, it will be even easier to cover them up because computer-based options will, by virtue of their source, carry great weight with many policy people and voters." [29]

The weight that such technical proposals carry has led some contemporary social critics to label our society "technocratic." In their use of the term, it seems to be a synonym for advanced technological society, rather than a description of the political decision-making process. Roszak, for example, says: "By the technocracy, I mean that social form in which an industrial society reaches the peak of its organizational integration. It is the ideal men usually have in mind when they speak of modernizing, updating, rationalizing, planning. . . . Understood . . . as the mature product of technological progress and the scientific ethos, the technocracy easily eludes all traditional political categories. Indeed, it is characteristic of the technocracy to render itself ideologically invisible. Its assumptions about reality and its values become as unobtrusively pervasive as the air we breathe." [30]

Often, then, the objections to "technocracy" are an expression of distaste for the ethos of a scientific age. Scientific method and technical rationality are seen as encroaching on humanitarian and democratic values. If a trend toward "meritocracy" may be seen as rooted in the American value dilemma of "equality" versus "achievement," so might a trend toward "technocracy" be related to the contradiction between two other American values: "democracy" and "secular rationality." As "secular rationality" is enhanced through the use of scientific and technical expertise in decision making and by the pervasiveness of a scientific mentality, some sacrifice of democratic values seems to be implied. Traditional forms of democratic participation and electoral consent become less meaningful, if "economic, military, and social policies become increasingly technical, long-range, machine-processed, information-based, and expert-dominated." [31] In the face of the complexities involved in many political decisions, the question of the proper balance between expert and popular authority is a most difficult one.

An understanding of the relationship between technical rationality and democratic politics requires an examination of the new techniques of governing, such as computerized information systems, systems analysis, and planning-programming-and-budgeting systems. These techniques are principal symbols of "technical rationality" and, as such, they evoke all the hopes and fears associated with it. The problems associated with these techniques may be seen as falling into two principal categories: the difficulties involved in implementation and meaningful use, and the challenges or dangers posed for democracy. The two are often related. For example, the dangers are sometimes exaggerated because of insufficient appreciation of the technical difficulties and political constraints involved in implementation of the new techniques.

Information Technology

The gathering and organization of information are obviously first steps for any of the new decision-making techniques, and the computer has been welcomed as a tool to expedite and improve information processing. Computers are often introduced into an organization to handle routine administrative, recordkeeping, and calculating functions; the intent is to automate the existing arrangements while "leaving organization structure and procedures more or less as they were before." Yet in the process, so much more information may be generated that "the question of who is to read and act upon it may provoke crises. For, traditionally, information is for men to read, to digest and act upon." [32]

Similarly, computers often compel managers and administrators to take explicit account of the distinction between the process of making decisions and the process of carrying them out. "Decisions to be made with the use of computers require that someone must determine first what information variables should be fed into the computer. To carry out the decisions afterward requires action on the processed information supplied by the computer. These two aspects of the decisions are now more likely to be handled by different people." [33] Such computer-generated pressures toward increased rationalization of organizational procedures also arise as a result of the analytical techniques that accompany computerized information systems. Systems analysis, for example, exerts a pressure on those who use it to explicate their goals and assumptions more clearly.

In the short run, however, these rationalizing forces tend to be overpowered by the political and organizational contexts in which computers are used. In a politicized environment like that of a government agency, considerations of power and influence tend to supersede the dictates of efficiency and systems analysis. The very collection of information may be problematic, because information is an important element in the maintenance of power and influence and is therefore a precious commodity that is often doled out parsimoniously and for the purpose of furthering the interests of the information holder. Government agencies tend to withhold information that puts them in an unfavorable light, and they often induce a certain fuzziness—if not distortion—in order to maintain their competitive positions in relation to other agencies. Moreover, while systems analysis begins with a long-range goal and then employs careful measurements and calculations to determine what the programs established to implement this goal are actually achieving, political considerations often dictate a focus on short-range goals, the "appearance" of effort, and a thin spreading of resources over many programs in order to gain support from diverse groups.[34]

Even in the somewhat less politicized environment of a business firm, the structure and doctrines of the organization often generate what Wilensky has called "information pathologies." These occur, for example, when an exceedingly hierarchical structure blocks upward communication of information, when excessive specialization and interdepartmental rivalry generate misleading information or prohibit the flow of information, when an overcentralized structure keeps top management too overloaded and out of touch, or when a doctrine of getting the "facts" to "fill in the gaps" separates the collection of facts from their interpretation and excludes experts from policy deliberations.[35]

In the absence of organizational change, the introduction of computerized information systems is not likely to overcome these pathologies. Indeed, "insofar as top managers ask the wrong questions and muster poor intelligence, wrong decisions will be more efficiently arrived at, and poor judgment, now buttressed by awesome statistics, will be made more effective. Wherever the new tools and perspectives are institutionalized, more weight will attach to data and systems analyses, whatever their quality. The chance is increased that information errors will ricochet at high speed throughout the system." It is only when "top managers know what questions to ask, and can elicit good data (i.e., overcome the structural and doctrinal roots of intelligence failure), [that] their decisions will be more efficient." [36]

After observing the information pathologies that beset the highest levels of national decision making, Oettinger points out that the combination of computers and new communications technology *can* provide a way of overcoming these deficiencies, *if* institutional change accompanies their introduction. Currently, communication of information to the "Chiefs," such as the president or department heads, is blocked or distorted by their "Indians," who often also lack access to information that is not within their immediate purview. Two types of information flow between Indians and Chiefs are possible. In the first, which is suitable for routine processes of government, the selection, collation, and interpretation of data is delegated entirely to the Indians. Filtering of information is maximum, information overload is minimized, and the scarce attention of the Chief is conserved. In the second, which is better suited to extraordinary crises, there is a minimum of filtering by the Indians, and the Chief gets a great deal of information. Although he may suffer from information overload, his options for action are far wider because he knows more. A given agency is likely to be organized exclusively around one of these extremes. But neither is workable alone; what is needed is a flexible system that can move toward either pole, depending on the situation. Oettinger recommends a system in which the Chiefs would have direct access to and control and supervision over all systems, so as to increase their ability to choose what information to receive themselves and what information to have their subordinates receive, while the Indians would have unfettered access to all systems in all hierarchies, so that analysts and experts could have interchange with people and data in both their own and other hierarchies.[37]

Although changes of this sort might improve the decision-making process, and although the new technology makes them practicable, inertia and the resistance of those who fear they would lose power in such a system are effective blockages. A scheme such as Oettinger's would not change the lines of authority within the government hierarchy, but it might—by broadening the access of information—decrease the power of those who now have exclusive access to the information. At all levels of government, this equation of information and power inhibits establishment and effective use of computerized information systems for decision-making purposes. The pooling of information from a variety of agencies or departments—which could provide a better data base and reduce costly duplications—sparks fears that the authority of the agencies in question, or specific persons within them, will be diminished.

Some consolidation of authority at the level of local government would appear to be required, for example, if one considers the fact that the United States today has 50 states, "3,043 counties, 17,144 towns and townships, 17,997 municipalities, 34,678 school districts, and 18,323 special districts. The number of agencies among these 91,236 areas apparently has never been recorded." [38] Although expanded use of computers will add to the pressures for consolidation at the local government level, in the near future this is more likely to take the form of consolidation of files than of consolidation of authority.[39]

Even this much will be difficult, since information systems have "power payoffs" as well as "technical payoffs." Higher-level officials and administrators tend to gain power over their lower-level counterparts, administrators generally tend to gain power over legislators, and technically sophisticated and well-organized groups tend to gain power over the less sophisticated and organized. This creates resistance to the introduction of comprehensive information systems precisely because these can lead to shifts in power and authority. At the same time, smaller-scale systems tailored to more specific administrative tasks tend to be accepted and installed more readily than the more comprehensive information systems, partly because their "technical payoffs" can be more easily demonstrated.[40]

In addition to such issues of power and resistance to organizational change, the introduction and implementation of information systems and systems analysis are held back by shortages of qualified personnel, inadequacies in computer hardware and software, insufficient understanding of currently unquantifiable information, the difficulties of deciding on public objectives and applying these techniques to complex decision making, and, perhaps above all, the dearth of demonstrable payoffs.

The efficacy of the techniques themselves—irrespective of their incompatibilities with political considerations—is far from proven. Whereas in the early days they were looked upon as a panacea for political, social, and administrative ills, a reaction of skepticism has now set in. Even if their practical utility remains unproven, however, politicians may find it useful to announce that they are using the most modern techniques, and hence the language rather than the spirit of systems analysis may be adopted. In this atmosphere, the benefits that might be derived from systems analysis are unlikely to be realized. As one observer has commented, "The political arena in which the game is played discourages rigorous review, since large sums of public money are involved. In their expenditure, everyone must look good. No official is so possessed of the death wish as to admit that the venture was anything but successful. Consequently, every aspect of the transaction, quite irrespective of its true color, comes through tinted with a glow of success. The mixture of salesmanship and politics may, ultimately, undermine the state of the art, for short-run, pervasive zeal for self-perpetuation practically guarantees stagnation. With little benefit of feedback from earlier experience, the same level of sophistication remains, with the same shortcomings, the same deficiencies, the same old excuses. Conceptual and methodological mutations are needed in order to create a tool useful in social planning, but these cannot occur unless there are open channels of inquiry and assessment free from public relations embellishments." [41]

However ineffective such systems may be as yet, the potential uses of large stores of computerized information have generated public anxiety. As a result, attempts to consolidate information at the federal level have thus far not borne fruit. Based on the premise that the existing federal statistical system is "too decentralized to function effectively and efficiently," [42] the Bureau of the Budget in 1965 proposed the establishment of a National Data Center. The proposal immediately became the subject of controversy. While proponents argued that a national data bank is sorely needed as an aid in public decision making, opponents evoked images of "1984." Although the intent of the proposal was to pool information that already exists in various government agencies rather than to establish dossiers on individual citizens, it was argued that information about individuals could be easily acquired after such a system was created. In response, it was maintained that both legal and technical safeguards could be built into the system to insure against invasions of privacy.

Invasions of privacy are a common occurrence, of course, even in the absence of a national data bank. Computerized information about individuals seems especially threatening, however, because it would probably make it more difficult than now for individuals to correct errors in or provide interpretations of their personal records. On the other hand, a centralized system might also decrease the risk of illegitimate leaks of information. In the longer term, moreover, widespread use of information systems may lead to changes in our conceptions of the nature of privacy. As Michael has said, "it would not be surprising if, in the future, people were willing to exchange some freedom and privacy in one area for other social gains or for personal conveniences." [43] And it may be that "we shall have a new measure of privacy: that part of one's life which is defined as unimportant (or especially important) simply because the computers cannot deal with it." [44]

Whatever the difficulties and time lags involved, most analysts seem to agree that the establishment of some sort of national data system is likely. The crucial issue thus becomes: Who will have access to the data? This question is important not only for the preservation of privacy, but also for the maintenance of a society in which all power does not reside in the government.

In the hope of averting exclusive government control of the information contained in data banks, some observers have suggested that citizens must have access to the data contained therein. Thus, it has been proposed that multiaccess computer terminals, all linked to the same data banks and computers as those used by the planning and governing agencies, could be placed around the city, and the computer programs and conceptual schemes, as well as the data, could be made known. Because the average citizen could not use this access effectively, specialists would have to be hired by citizen groups to interpret the information for them.[45] Alternatively, it has been suggested that a national data center should be housed outside the executive branch of government. All branches of government as well as the citizenry should have access to it, and both private and public organizations should be able to contribute to it. Such a scheme could have high costs in loss of efficiency, because congressional or citizen groups might subject federal planners to intense scrutiny and criticism, so that they would be required to spend large amounts of time and money in refuting criticisms and justifying policies. But, it is argued, this may be worth the cost.[46]

Making trade-offs of this kind between the enhanced efficiency that computers might bring to government decision making on the one hand and the rights of the citizenry on the other is one approach to the problem of possible infringements on democracy. Another approach that has been suggested is to use computer technology to bolster the role of the citizenry in decision making, through a system of "participatory democracy" or "citizen feedback." Computer technology makes it possible to institute a system of "instant voting" in which the electorate could register their opinions on various issues. In the extreme form of "participatory democracy," the results of such votes would be binding on the decision makers.

In a more moderate version, citizens would use computer technology to provide politicians and decision makers with information about their views. Such feedback systems, it is argued, would redress the current imbalance in the flow of information between the government and the citizenry and would relieve some of the frustrations currently expressed with the workings of the ordinary channels of government.[47] The latter is a questionable assumption, since any failure of government to act in accordance with the known wishes of certain citizen groups would be likely to increase popular frustration. The expectation that political actors would respond directly to the wishes of their constituents might also work havoc with the representational system. It has been pointed out that because the representational process involves the reconciliation of conflicting interests, it is essential to the system as a whole that representatives can choose to ignore or selectively interpret some of the wishes of their constituents.[48]

Accompanying concerns about maintaining access to information stored in data banks is the fear that the practitioners of systems analysis and allied techniques will dictate social values and policies. This fear may have several sources. Perhaps the simplest one is the assumption that "the systems boys" will take over. As noted earlier, there seems to be little indication that this will happen.

On a somewhat higher level of abstraction, the fear has been expressed that use of these techniques may transform them from means to ends; that is, that so much emphasis will be placed on the measurable and quantifiable that the intangibles of human values will be ignored or subordinated. Most observers have noted that systems analysis and allied techniques cannot deal with value problems or problems of irreconcilable objectives. Yet it is feared that they will be used even in such inappropriate situations. If this happens, the values of efficiency and measurable output will become ends rather than means. Thus, "one programmer may value cheap, efficient roads and may ask the computer to provide him with specifications for such roads, whereas another may value expensive, beautiful roads. In this second instance, the values are less clear-cut and more difficult to measure: What is beautiful? How much should be spent for how much beauty? Not only are such parameters difficult to define and measure, they may in some instances be difficult to admit or recognize. . . . [There is a danger, therefore, that] complex decision systems involving human parameters will be broken down into routine segments which are more or less independent of human reaction, and that the combination will then be called a credible simulation of the total system. Such a danger has always existed in all categories of problem solution; however, with the advent of increasingly effective computers, the danger is becoming more seductive and more far-reaching in the scope of its influence." [49]

Even though this danger should not be dismissed, it should also not be exaggerated. Ample weight must be given to the power of specifically political factors. Both systems analysis and PPB are basically management tools. They are used to discover how we can achieve a given end most effectively and efficiently with the resources that are available. The ends to be achieved are not set by the analyst or programmers themselves. They are set by the decision makers and are thus subject to all the constraints under which political leaders and administrators labor. If the technical analysts give insufficient weight to nonquantifiable values and interests, their proposals are likely to be politically unsalable. There is nothing automatic about the implementation of their recommendations, therefore. In fact, such recommendations are often left to gather dust because the decision makers responsible for implementing them find them unworkable.[50]

The uses of information stored in government data banks might pose problems for government as well as for the citizenry. In some cases, the simple fact of gathering information may be politically difficult because it might raise the expectation that something will be done about the phenomenon or problem at issue. Hence, "it is wisdom, not cynicism, to urge caution in extending diagnostic measures of social phenomena beyond the system's capacity to respond to the problems which are unveiled. While it is necessary to illuminate problems for planning purposes and to stimulate the requisite actions, such illumination can also produce trouble and disillusionment." [51] In a more general sense, "information systems may generate knowledge which undercuts traditional governmental justifications for inaction, and this same knowledge can be used by dissenting groups to intensify political pressure for a redistribution of societal resources." [52]

Technology Assessment

Although the use of information technology in government is still in its infancy, the desire for more effective use of knowledge and technology is burgeoning. There is a growing sentiment that while the United States has been spectacularly successful in developing new technologies, it has been much less successful in controlling their effects and putting them to use in the solution of social problems. The mechanisms that support technological growth—R&D contracting, executive agencies and congressional committees, and what has been called the military–industrial complex—have, by and large, performed effectively in the task of promoting technological development. But, as we have noted, this has often been done at the cost of other social goals and of frequent failure to provide adequate control over the uses of technology.

The movement to establish some form of "technology assessment" within the government is a response to this failure. Like all pressures in this direction, technology assessment is fraught with the difficulties of inadequate knowledge, insufficient ideological legitimation, and the resistance of vested interests. The search for mechanisms to overcome these difficulties is under way.

In 1967, Congressman Emilio Daddario proposed that a Technology Assessment Board be established to advise Congress on the social costs and benefits of technology.[53] The board would assess the desirable and undesirable effects of technological programs and offer policy recommendations. There are many difficulties in such a procedure: methodological problems of how to foresee the social consequences of a technology, value problems of how to weigh the various consequences against each other and how to evaluate the short- as against the long-run impacts and implications.

There are also questions about the political utility and viability of a system of technology assessment. Because a new governmental institution for technology assessment might be ineffective owing to the vested interests that arise in connection with technological developments, an alternative proposal has been put forth to create a board that would play the role of "devil's advocate," by calling attention to the negative consequences of new technology, and by functioning as a lobby to press for their containment.[54] Since a board of this kind would have a purely advisory role, however, it is unclear how much impact it would have, although it is possible that some institutionalized form of constructive muckraking might be effective.

Yet another form of institutionalization of technology assessment has been suggested in a report of the National Academy of Sciences. It proposes that we establish "not a single assessment mechanism but a network of such mechanisms extending throughout government and the private sector" and that the "evidence and arguments on which major decisions are based" should be "open to public scrutiny" and "subject to timely review in appropriate public hearings." [55]

Probably more public attention has been devoted recently to countering or controlling the negative effects of technology than to making more effective use of it. One particular way of using technology for social benefit that has been suggested by Weinberg is the development of "technological fixes" for the solution of social problems.[56] While the idea of "technological fixes"—subsequently called "shortcuts" by Etzioni[57]—is not as much "in good currency" as is that of technology assessment, it is not entirely without supporters. The argument is that technological solutions to social problems are often easier to effect than political or economic solutions. Although they may not eliminate the underlying problems, technological fixes can serve as ameliorative devices and allow us to buy the time necessary to get at the cause of the social problem. Examples of technological solutions to social problems include (1) the production of electric cars, which might provide an easier solution to the pollution problem than legislation to enforce the use of antipollution devices, and (2) production of a safe cigarette, which is an easier solution to the smoking problem than exhortations to break the habit. Weinberg has argued that the skills of such large laboratories as Argonne and Oak Ridge should be redeployed for work on such technological solutions to social problems.

Whether these proposals for technology assessment or technological fixes will become meaningful parts of a transformed political structure remains to be determined. In any case, the cumulative effects—direct and indirect—of technological change on the polity are likely to become more important and to exert increased pressure for more public control over, and public guidance of, technological development.

□ □ □

1. Martin Shubik, "Information, Rationality, and Free Choice in a Future Democratic Society," *Daedalus*, 96 (Summer 1967), 778 and 772.
2. Daniel Bell, "The Dispossessed," in Bell, ed., *The Radical Right* (New York: Anchor Books, 1964), pp. 19–20.
3. Victor C. Ferkiss, *Technological Man: The Myth and the Reality* (New York: George Braziller, Inc., 1969), pp. 190–91.
4. Shubik, *op. cit.*, p. 776.

5. Peter F. Drucker, *The Age of Discontinuity* (New York: Harper & Row, Publishers, 1969).
6. Michael Harrington, *The Accidental Century* (Baltimore: Penguin Books, Inc., 1967).
7. Henry S. Kariel, *The Promise of Politics* (Englewood Cliffs, N.J.: Prentice-Hall, Inc., 1966).
8. Herman Kahn and Anthony J. Wiener, "Faustian Powers and Human Choices: Some Twenty-First Century Technological and Economic Issues," in William R. Ewald, Jr., ed., *Environment and Change, The Next Fifty Years* (Bloomington, Ind.: Indiana University Press, 1968), pp. 101–31.
9. Ferkiss, *op. cit.*
10. Daniel Greenberg, "The Myth of the Scientific Elite," *The Public Interest*, 1 (Fall 1965), 51–62.
11. Jean Meynaud, *Technocracy* (London: Faber and Faber, 1968).
12. Robert Boguslaw, *The New Utopians: A Study of System Design and Social Change* (Englewood Cliffs, N.J.: Prentice-Hall, Inc., 1965).
13. James R. Schlesinger, "Systems Analysis and the Political Process," *Journal of Law and Economics*, 11 (October 1968), 281–98.
14. Samuel P. Huntington, "Political Modernization: America vs. Europe," in Reinhard Bendix, ed., *State and Society: A Reader in Comparative Political Sociology* (Boston: Little, Brown and Company, 1968), pp. 178–79.
15. Daniel Bell, "Notes on the Post-Industrial Society (II)," *The Public Interest*, 7 (Spring 1967), 102–18.
16. Sidney Verba, "Democratic Participation," in Bertram M. Gross, ed., "Social Goals and Indicators for American Society: Vol. II," *Annals of the American Academy of Political and Social Science*, 373 (September 1967), 55.
17. Emmanuel G. Mesthene, *Technological Change: Its Impact on Man and Society* (Cambridge, Mass.: Harvard University Press, 1970), p. 55.

18. *Ibid.*, p. 74.
19. Drucker, *The Age of Discontinuity*, pp. 234 and 236.
20. Mesthene, *op. cit.*, pp. 40–41.
21. Milton Katz, "The Function of Tort Liability in Technology Assessment," *Cincinnati Law Review*, 38 (Fall 1969), 587–662.
22. See Seymour Martin Lipset, "A Changing American Character?" in Seymour Martin Lipset and Leo Lowenthal, eds., *Culture and Social Character* (New York: The Free Press, 1961), pp. 136–71.
23. Zbigniew Brzezinski, "The American Transition," *The New Republic*, 157 (December 23, 1967), 18–21.
24. John McDermott, "Technology: The Opiate of the Intellectuals," *The New York Review of Books*, 13 (July 31, 1969), 33.
25. H. L. Nieburg, *In the Name of Science* (Chicago: Quadrangle Books, Inc., 1966), pp. 132–33.
26. Melvin M. Webber, "The Politics of Information," in Donald N. Michael, ed., *The Future Society* (New Brunswick, N.J.: Trans-action Books, 1970), pp. 14–15.
27. *Ibid.*, pp. 16–17.
28. Alain Enthoven, "Discussion," in Martin Greenberger, ed., *Computers, Communications and the Public Interest* (Baltimore: Johns Hopkins Press, 1971), pp. 99–101.
29. Donald N. Michael, "On Coping with Complexity: Planning and Politics," *Daedalus*, 97 (Fall 1968), 1183.
30. Theodore Roszak, *The Making of a Counter Culture: Reflections on the Technocratic Society and Its Youthful Opposition* (New York: Doubleday & Company, Inc., 1969), pp. 5 and 8.
31. Emmanuel G. Mesthene, "How Technology Will Shape the Future," *Science*, 161 (July 12, 1968), 142.
32. H. A. Rhee, *Office Automation in Social Perspective* (Oxford, England: Basil Blackwell, 1968), p. 75.

33. *Ibid.*, p. 95.
34. James R. Schlesinger, "Systems Analysis," *loc. cit.*
35. Harold L. Wilensky, *Organizational Intelligence* (New York: Basic Books, Inc., Publishers, 1967).
36. *Ibid.*, p. 185.
37. Anthony G. Oettinger, "Compunications [*sic.*] in the National Decision Making Process," in Greenberger, ed., *Computers, Communications and the Public Interest*, pp. 74–91.
38. Paul Armer, "Computer Aspects of Technological Change, Automation, and Economic Progress," *The Outlook for Technological Change and Employment*, Appendix Vol. I to National Commission on Technology, Automation and Economic Progress, *Technology and the American Economy* (Washington, D.C.: U.S. Government Printing Office, 1966), p. 221.
39. *Ibid.*, p. 222.
40. Anthony Downs, "A Realistic Look at the Final Payoffs from Urban Data Systems," *Public Administration Review*, 27 (September 1967), 204–10.
41. Ida Hoos, "A Realistic Look at the Systems Approach to Social Problems," *Datamation*, 15 (February 1969), 228.
42. Carl Kaysen, "Data Banks and Dossiers," *The Public Interest*, 7 (Spring 1967), 53.
43. Donald N. Michael, "Speculations on the Relation of the Computer to Individual Freedom and the Right to Privacy," *The George Washington Law Review*, 33 (October 1964), 273.
44. *Ibid.*, p. 279.
45. Donald N. Michael, "On Coping with Complexity: Planning and Politics," *Daedalus*, 97 (Fall 1968), 1179–93.
46. Robert O. MacBride, *The Automated State: Computer Systems as a New Force in Society* (Philadelphia: Chilton Book Company, 1967), pp. 169–87.
47. Chandler H. Stevens, "Citizen Feedback: The Need and the Response," *Technology Review*, 73 (January 1971), 39–45.

48. Heinz Eulau, "Some Potential Effects of the Information Utility on Political Decision-Makers and the Role of the Representative," in Harold Sackman and Norman Nie, eds., *The Information Utility and Social Choice* (Montvale, N.J.: AFIPS Press, 1970), pp. 187–99.
49. David L. Johnson and Arthur L. Kobler, "The Man–Computer Relationship," *Science*, 138 (November 23, 1962), 875–76.
50. See Ida Hoos, "A Critique of the Application of Systems Analysis to Social Problems," paper given at the 13th Annual Meeting, American Astronautical Society, Dallas, Texas, May 1–3, 1967; and Harold R. Walt, "The Four Aerospace Contracts: A Review of the California Experience," *Applying Technology to Unmet Needs*, Appendix Volume V to *Technology and the American Economy*, *op. cit.*, pp. 47–53.
51. Raymond Bauer, "Societal Feedback," *Annals of the American Academy of Political and Social Science*, 373 (September 1967), 190–91.
52. Alan F. Westin, "Information Systems and Political Decision-Making," in Irene Taviss, ed., *The Computer Impact* (Englewood Cliffs, N.J.: Prentice-Hall, Inc., 1970), p. 144.
53. Emilio Q. Daddario, "Technology Assessment," Committee Print, U.S. House of Representatives, Committee on Science and Astronautics (Washington, D.C.: U.S. Government Printing Office, 1967).
54. Harold P. Green, "Technology Assessment and the Law: Introduction and Perspective," *The George Washington Law Review*, 36 (July 1968), 1033–43; and Michael Wollan, "Controlling the Potential Hazards of Government-Sponsored Technology," *The George Washington Law Review*, 36 (July 1968), 1105–37.
55. U.S. House of Representatives Committee on Science and Astronautics, *Technology: Processes of Assessment and Choice*, Report of the National Academy of Sciences (Washington, D.C.: U.S. Government Printing Office, July 1969), pp. 77 and 67.

56. Alvin M. Weinberg, "Social Problems and National Socio-Technical Institutes," in *Applied Science and Technological Progress*, A Report to the Committee on Science and Astronautics, U.S. House of Representatives, by the National Academy of Sciences (Washington, D.C.: U.S. Government Printing Office, 1967), pp. 415–34.

57. Amitai Etzioni, " 'Shortcuts' to Social Change?" *The Public Interest*, 12 (Summer 1968), 40–51.

TECHNOLOGY and VALUES

The issues that technology poses for governance are intimately related to value questions, since political change requires accompanying value changes. If the governance of a complex society cannot remain the same as it was when the social structure was simpler and technology less prominent, neither can the values of that society remain unchanged. The enhancement of government power, the trend toward planning, and their underlying social causes have important consequences for our values.

The dominant American values of individualism, achievement, success, progress, efficiency, practicality, and rationality have been eminently suited to the development and application of technology. But technological change, in turn, results in value change—either indirectly through its effects on social structure or more directly by changing the costs and benefits associated with the realization of different values.

The study of value change is fraught with difficulty, however. First of all, as conceptions of the desirable that guide action, values represent factors that underlie behavior and attitudes.

Hence, it is difficult to construct meaningful indicators with which to measure or study them. Second, since values are the most deeply rooted aspects of a culture, they tend to change more slowly than technologies, organizations, or specific behaviors. The process of value change is thus a gradual one in which it is difficult to determine at what point the various disaffections from existing values amount to significant change. Further complicating the problem is the fact that in a complex society, there are various subcultures that maintain values that differ from those of the dominant culture. The significance of such subcultures and the extent of their influence are often difficult to measure.

THE RELATIONSHIP BETWEEN TECHNOLOGICAL CHANGE AND VALUE CHANGE

The mechanisms by which technological change effects value change may be either direct or indirect. The direct effects of technology on values occur as a result of the capacity of new technologies for creating new opportunities. "By making available new options, new technology can, and generally will, lead to a restructuring of the hierarchy of values, either by providing the means for bringing previously unattainable ideals within the realm of choice and therefore of realizable values, or by altering the relative ease with which different values can be implemented —that is, by changing the costs associated with realizing them." [1]

In this way, technological change impinges on the intrinsic sources of change or instability within value systems. Two basic instabilities in all values systems are the inherent contradictions between diverse values within any given value system (or the contradictions involved in implementing them) and the tension between principles and practices (the "ideal–real" gap). By altering the relative costs of implementing them, technology may alter the balance among competing values. Thus, as noted earlier, technological change has exacerbated the long-standing conflict between "equality" and "achievement" by altering the costs associated with them. Also, by making some heretofore unrealizable ("ideal") values realizable ("real"), technology may "call our bluff" and thereby create new tensions and perhaps ultimately a redefinition of values. Such tensions are currently being experienced in regard to "equality," for example, as our technology-based affluence offers the possibiilty of implementing our proclaimed ideals of social and economic equality. Similarly, sophisticated computer technology increasingly makes possible the rapid and efficient collection of the opinions of the citizenry, and thus raises the possibility of instituting the ideal of direct democracy that is reflected in mythologies of "New England town democracy." As long as such values remained impossible to achieve, they were not seriously problematic. As technology makes them attainable, the disagreements among different segments of our population become more open and manifest.

The changes that technology brings in the costs and benefits involved in realizing diverse values have been used as a basis for predicting value change. Thus, Nicholas Rescher has constructed a typology of "modes of upgrading" and corresponding "modes of downgrading" of existing values. Whether a particular value is likely to be upgraded or downgraded in the face of social and technological change is determined by means of a cost-benefit analysis. Social and technological change, Rescher argues, "can alter the costs involved in realizing a value (either downwards, as with air-travel 'safety' in recent years, or upwards, as with urban 'privacy'). On the other hand, such changes in the life environment can also alter the benefits derivable from realizing a value (as e.g., the benefits to be derived from wealth decline in an affluent society)." [2] On this basis Rescher predicts, for example, that nationalistic values will be downgraded because the cost of maintaining them will be too high, while social accountability will be upgraded and self-advancement downgraded as automation, "socioeconomic rigidity," and the size and complexity of the social system are extended.

The framework used here is essentially a combination of extrapolation and analytical forecasting based on cost-benefit techniques. The idea that technological and other social changes have an effect on values by altering the costs and benefits attached to their realization clearly offers an analytic handle with which to study the relationship between technological and value changes. But its application rests on a high degree of intuition, since the hierarchy of current values—and their attendant costs and benefits—is never spelled out precisely. Thus, it is not always clear why one value is seen to be upgraded in the face of threat (or increased cost of maintaining or realizing it), while another value so threatened is seen as being downgraded. Moreover, the strength of adherence to a particular value today is often taken as the central determinant of the strength of this value in the changed social situation of the future. Hence the model fails to take into account that kind of value change in which values that were earlier of great importance become downgraded, and vice versa.

In the study of values, however difficult, the analyst has at his disposal records of behavior, survey research materials, laws, folklore, mass media, arts, and other public documents. He can examine such materials at different time periods in the attempt to discern value changes. Such techniques are not applicable to the study of future values. To predict value change, therefore, one might start with the assumption that values change in response to changed social conditions. Then, if changes in social conditions can be forecast, one can attempt to predict the value changes that are likely to result. Aside from the difficulties in making accurate predictions about future social conditions, there are two serious problems in this type of procedure. First, in the attempt to extrapolate value change directly from other social changes, one is likely to arrive at inaccurate conclusions because of a failure to take into account the time lags in the process of social change. Second, if one tries to avoid this pitfall by making some assumptions about which values are likely to persist despite changed social conditions and which are more likely to change, there is a danger of imputing values to future generations on the basis of the strength with which various values are adhered to today. To overcome both these pitfalls, one could adopt a procedure of examining the present inconsistencies between social requirements and traditional values and noting those values that appear to be under stress or in need of change. One might then speculate about the nature and rapidity of such value change. Despite the difficulties in this approach and the tentativeness of any conclusions that might emerge, it might have the merit of shedding light on current value patterns.[3]

Technological and social change have produced two principal strains in the contemporary American value system. The first is an apparent need for adjustment in the individualistic pattern of values. The second is a profound questioning of the nature of values and value commitments. In both cases, the effects of technological change are indirect.

INDIVIDUALISTIC VERSUS COLLECTIVE ORIENTATIONS

Despite the important ways in which values determine the technologies that are developed and applied,[4] in recent years the lion's share of attention has been devoted to the ways in which technological changes impinge on values and social structure. It may well be, as Robert Heilbroner suggests, that a position of technological determinism has been prevalent in Western society since 1700 because in this historic epoch "the forces of technical changes have been unleashed, but . . . the agencies for the control or guidance of technology are still rudimentary." The future, he argues, may be different; "the pressures in the future will be toward a society marked by a much greater degree of organization and deliberate control." [5]

As noted earlier, the trend toward greater social planning raises some important value issues. While the governing of any nation or institution requires an orientation toward the welfare of the whole and imposes some constraints upon the behavior of individuals and groups within it, a system of social planning strengthens these needs and makes them more visible. Planning thus imposes a strain on the strongly individualistic values of American society. Americans have cherished the right of individuals and individual groups to act in accordance with their own self-interest, and have assumed that the net result for society must be beneficial. This assumption is no longer adequate.

As Garrett Hardin has so forcefully argued, the operation of "the commons" in which each rational herdsman decides that adding another animal to graze is more to his personal benefit than to the collective detriment of the commons, can no longer be tolerated; "freedom in a commons brings ruin to all." What was moral a hundred years ago, he points out—for example, the disposal of waste into the local river—is today unfeasible, because increased population density means that the accumulation of such singular acts results in an overload.[6] A central value problem for American society, then, is to disassociate the concept of freedom from laissez-faire conceptions of the free reign of individual interests.

A related, but more difficult, problem is to alter American values in the direction of the more collective orientation that appears to be necessary today. The American value system places little emphasis on the goals of the society as a whole. "Except for situations of national emergency, system-goal attainment [comes] last [in the American value hierarchy]; this . . . is primarily what we mean by our individualism." [7]

A major difficulty appears to reside in the fact that modern society has become, at one and the same time, both more individualist and more interdependent. Georg Simmel captured the essence of this dilemma at the turn of the century: "The deepest problems of modern life derive from the claim of the individual to preserve the autonomy and individuality of his existence in the face of overwhelming social forces, of historical heritage, of external culture, and of the technique of life. . . . The eighteenth century called upon man to free himself of all the historical bonds in the state and in religion, in morals and in economics. . . . In addition to more liberty, the nineteenth century demanded the functional specialization of man and his work; this specialization makes one individual incomparable to another and each of them indispensable to the highest possible extent. However, this specialization makes each man the more directly dependent upon the supplementary activities of all others." [8] Moreover, in the nineteenth century, "another ideal arose: individuals liberated from historical bonds now wished to distinguish themselves from one another." [9]

In the years since Simmel wrote this essay, social and technological changes have intensified both individualism and interdependence. High population densities and large-scale technologies enlarge interdependence, as pollution, blackouts, and other such difficulties amply demonstrate. Concurrently, greater social complexity and increased mobility make it possible for the individual to have more freedom from "historical bonds."

Because the actions of individuals in a complex technological society have more and greater ramifications for their fellows and society at large than they had in a simpler society, greater social control seems to be needed. At the same time, increased individuation would seem to imply that there is less intrinsic or natural social control today than earlier. What appears to be required, therefore, is the development of a value orientation in which individuals would "internalize" the value of social welfare, so that considerations of the public or collective welfare would play a larger role in individual behavior and greater effort would be expended in collective actions.

In the absence of such change, actions taken for reasons of self-interest often bring negative social consequences that, because of interdependence, also have deleterious consequences for the individual himself in the long run (as Hardin's example of "the commons" illustrates). Individuals also continue to expend little energy in collective actions, although the efficacy of purely individual action has been reduced. While the Good Samaritan continues to provide a model for behavior, it is no longer appropriate. The Good Samaritan depends upon his own resources to remedy a wrong; but the issue today is one of public resources and social organization. The moral agent has become the public agent, and the individual must show his skills in corporate and social action.[10]

The need for a value change of this sort is argued even by some critics who reject many of the institutional and cultural forms of advanced technological society. Thus, "Consciousness III," in its most general contours, demands both a high degree of individualism and a strong commitment to social welfare. It begins with the primacy of the individual and goes on to proclaim the importance of social needs. As Reich expresses it, "The initial premise of self leads not only to a critique of society, it also leads, in many representatives of Consciousness III, to a deep personal commitment to the welfare of the community." [11] The key element in this formulation is "full personality responsibility" for the social welfare.

An acceptance of the need for a more collective orientation and for social planning to promote it would not resolve the value problems, however, since the difficulties of determining what constitutes social welfare remain. Disagreements about its definition stem from differences in values and preferences and from disputes that rest upon an inadequate knowledge of the social consequences that would follow from a given course of social action. Although it was fashionable in the 1950's and 1960's to speak of an "end of ideology" in advanced "knowledgeable" societies, this picture would appear to be an incomplete one. It cannot be denied that knowledge plays a much more important role in government and politics today than in the past. But can the following assertion be supported? "If one thinks of a domain of 'pure politics' where decisions are determined by calculations of influence, power, or electoral advantage, and a domain of 'pure knowledge' where decisions are determined by calculations of how to implement agreed-upon values with rationality and efficiency, . . . the political domain is shrinking and the knowledge is growing." [12] There can be little doubt that where political differences are based solely on ignorance of outcomes, greater knowledge will lead to greater consensus. But in cases where differences are based upon conflicting interests or values, knowledge might enlarge the conflict by providing greater certainty about the consequences of a given policy action.

Planning sharpens value conflicts and inconsistencies because it requires a choice between competing values as these apply to specific actions and decisions. Hence, educational planning must cope with dilemmas of fostering equality or achievement, industrial planning faces choices of productivity versus leisure, and so on. The rationalization of values that is implied in social planning is not easily handled, particularly in a period of rising expectations and the concomitant airing of demands from diverse social groups.

VALUE UNCERTAINTIES

At the present time, the problem of value inconsistencies is submerged within the larger problem of uncertainty concerning values. The cultural underpinnings that secure commitments to a given set of values seem to have become less firm. There is a sense of the relativity of all values, of the absence of any "eternal verities." Values are often adhered to in a flexible way, so that commitments to them are weak and traditional values are readily and frequently questioned. It is this dimension of value change that is at the root of the sense of "value crisis" in our age.

The factors responsible for this uncertainty may be traced to the social and cultural changes that have resulted from technological change. The technologies that have brought forth a densely populated urban society in which physical distance is less relevant and in which heterogeneous groups are brought together readily may be seen as one factor in the development of value uncertainty. "Population growth, urbanization, and the developments in communication, transportation, and the mass media all serve to increase contact and interaction between diverse peoples and groups. Under conditions in which social groups maintaining differing values are relatively insulated from one another, it is easy to maintain the group identity or value system, with perhaps only minimal consciousness thereof. However, as contact with representatives of other value patterns becomes more frequent and routine, a greater consciousness of values emerges which may bring with it a degree of value relativism." [13]

Such value relativism is further stimulated by mobility, role complexity, and the scientific or cognitive orientation of our culture. Widespread mobility and a perception of constant social change help to feed value uncertainties because the absence of a fixed environment and the consequent need to adapt to a series of different behavior standards make it difficult to maintain allegiance to a fixed set of values. The complexities involved in playing the multiple roles required of individuals in modern societies make the individual's adaptation to society a far more conscious process than it once was. This consciousness of social processes feeds the sense of uncertainty about values by reducing the scope of what can be "taken for granted."

The analytic stance toward society and values is further enhanced by the scientific world view and the application of scientific knowledge. The characteristics of science may be said to have infiltrated into the culture at large. Even though such scientific attitudes as a "respect for evidence and a willingness to follow evidence wherever it leads" [14] could probably never prevail in pure form, there has been an increasing tendency to look toward science and the scientific method for legitimation. In most spheres of action, evidence is collected and weighed and the "search for global solutions or global generalizations" [15] is abandoned. As a result, we may be faced with the irony that although meaning- and purpose-oriented explanations of the universe may be more needed today than in the less complex societies of the past, the characteristic attitude of science in "eschewing the more fundamental, the more 'metaphysical' issues" [16] may have become sufficiently pervasive as to render the acceptance of such explanations unlikely.

This cognitive orientation has a profound effect on values. For "a society in which the store of knowledge concerning the consequences of action is large and is rapidly increasing is a society in which received norms and their 'justifying' values will be increasingly subjected to questioning and reformulation." [17] In the mass media, the "repeated re-examinations . . . of fundamental questions about religion, ethics, crime, etcetera" also take their toll. "The distinction between good and bad, right and wrong, between virtue and vice, once held to be self-evident by our forebears are blurred and reblurred." [18]

Observers have difficulty in foreseeing what the outcome of the current malaise will be. Aside from a continuation of the present tendencies, two alternative possibilities have been noted: that some new forms of commitment will emerge or that a new cultural pattern will develop that is based on a rejection of commitments as we have known them. The former view is espoused by advocates of the "counterculture." The latter view has been argued by Philip Rieff, who extrapolates from the current trends to a new "dystopia" in which culture itself is replaced by "a new anti-culture [which] aims merely at an eternal interim ethic of release from the inherited controls." [19] In Rieff's schema, "every culture has two main functions: (1) to organize the moral demands men make upon themselves into a system of symbols that make men intelligible and trustworthy to each other, thus rendering also the world intelligible and trustworthy; (2) to organize the expressive remissions by which men release themselves, in some degree, from the strain of conforming to the controlling [symbol system]. The process by which a culture changes at its profoundest level may be traced in the shifting balance of controls and releases which constitute a system of moral demands. . . . What appears now fairly certain is that the control-release system inherited from an older, mainly agrarian, culture into our technologically advanced, urban one cannot renew itself. . . . The religious psychologies of release and the social technologies of affluence do not go beyond release and affluence to a fresh imposition of restrictive demands. This describes, in a sentence, the cultural revolution of our time." [20]

Most social analysts would probably not accept Rieff's view. The idea that a culture could sustain itself indefinitely without a shared system of values and restrictions to which its members are committed is antithetical to the basic tenets of the social and psychological sciences. Kahn and Wiener take a position somewhat similar to Rieff's, but far less drastic. They argue that the decrease in demands made upon the individual as economic security increases will result in the rise of alienation. Conceptions of self will become more problematic, and the influence of traditional religion as well as the ideals and expectations of American democracy will continue to wane. "The most serious issue raised by these speculations," they suggest, "is whether they are not just modern manifestations of traditional 'aberrant' behavior, or whether they represent a reasonable adjustment or transition state to new traditions and mores." [21]

But what the nature of these new traditions and mores might be, if they are forthcoming, is all but impossible to predict. Some commentators have suggested that we try to manufacture the new values and goals; that is, to set about making innovations in this sphere deliberately. Thus, theologians have argued that traditional religion must self-consciously set about altering some of its assumptions that no longer fit or serve the modern world. Donald Schon has argued that since neither "a call to Revolt" nor "a call to Return" will solve the problems generated by a rapidly changing technology and society, a new ethic of change is needed. "The crucial form of experiment demanded by the technological, institutional and normative changes of our time is experiment in norms and objectives." [22] Such experiments would be guided by the values that underlie science and the process of discovery. Rejecting the ethos of science and technology, the "counterculture" suggests the development of a new style of life based on intuition and communion. Let us turn now to an examination of the effects of such value uncertainties on the psychology of individuals.

□ □ □

1. Emmanuel G. Mesthene, "How Technology Will Shape the Future," *Science*, 161 (July 12, 1968), 137.
2. Nicholas Rescher, "What is Value Change?" in Kurt Baier and Nicholas Rescher, eds., *Values and the Future: The Impact of Technological Change on American Values* (New York: The Free Press, 1969), p. 80.
3. For a more extended discussion of the relationship between

futurology and the study of values, see Irene Taviss, "Futurology and the Problem of Values," *International Social Science Journal*, 21 (1969), 574–84.

4. See, for example, Joseph Needham, "Science and Society in East and West," in Maurice Goldsmith and Alan Mackay, eds., *The Science of Science* (Harmondsworth, England: Pelican Books, 1966), pp. 159–88, for a discussion of the ways in which the value system and social structure of ancient China inhibited the development of technology.
5. Robert Heilbroner, "Do Machines Make History?" *Technology and Culture*, 8 (July 1967), 345.
6. Garrett Hardin, "The Tragedy of the Commons," *Science*, 162 (December 13, 1968), 1243–48.
7. Talcott Parsons, "A Revised Analytical Approach to the Theory of Social Stratification," in Reinhard Bendix and Seymour Martin Lipset, eds., *Class, Status and Power* (New York: The Free Press, 1953), p. 106.
8. Georg Simmel, "The Metropolis and Mental Life," in Kurt H. Wolff, ed. and trans., *The Sociology of Georg Simmel* (New York: The Free Press, 1964), p. 409. The essay was written in 1902–3.
9. *Ibid.*, p. 423.
10. Thomas F. Green, *Work, Leisure, and the American Schools* (New York: Random House, Inc., 1968), pp. 119–20.
11. Charles A. Reich, *The Greening of America* (New York: Random House, Inc., 1970), p. 230.
12. Robert E. Lane, "The Decline of Politics and Ideology in a Knowledgeable Society," *American Sociological Review*, 31 (October 1966), 657–58.
13. Irene Taviss, "Changes in the Form of Alienation: The 1900's vs. The 1950's," *American Sociological Review*, 34 (February 1969), 54–55.
14. Harvey Brooks, "Scientific Concepts and Cultural Change," *Daedalus*, 94 (Winter 1965), 66.
15. *Ibid.*, p. 68.

16. Ibid., p. 66.
17. Robin M. Williams, Jr., "Individual and Group Values," *Annals of the American Academy of Political and Social Science*, 371 (May 1967), 30.
18. Ezra J. Mishan, *The Costs of Economic Growth* (New York: Praeger Publishers, Inc., 1967), p. 132.
19. Philip Rieff, *The Triumph of the Therapeutic* (New York: Harper & Row, Publishers, 1966), p. 23.
20. *Ibid.*, pp. 232–33, 239.
21. Herman Kahn and Anthony J. Wiener, *The Year 2000* (New York: The Macmillan Company, 1967), p. 220.
22. Donald A. Schon, *Technology and Change* (New York: Dell Publishing Co., Inc., 1967), p. 216.

TECHNOLOGY and the INDIVIDUAL

While the life-styles of individuals have clearly been affected by the advent of such modern technologies as cars, jets, and televisions, such general social changes as the value uncertainties discussed in the preceding chapter also have important consequences for individuals. Because these are less easily discernible, however, there is much room for disagreement about the state of man's psyche in an advanced technological society.

In the days of the Industrial Revolution, an examination of the effects of technology on the individual would most probably have discussed the poor living conditions that were being imposed on the masses. Today, the focus of most discussions of technology and the individual tends to be on man's inner life. An advanced technological society, whether it is seen as increasing man's freedom or decreasing it, extending his sense of alienation or reducing it, seems to make individual adjustment a more complex process. As some writers view it, this is due to change, or breakdown, in man's traditional institutions: his family, his church, his government. Others attribute modern man's psychological difficulties to the values and cultural patterns of a highly technological and rapidly changing society. A reading of the voluminous literature on alienation would suggest that the extent and depth of man's disaffection with his society or his self are not really known. Nor is it clear whether there are any differences in this regard between modern man and his predecessors. Nevertheless, it would be hard to imagine that technological change and the social changes it brings could leave the individual unaffected.

The difficulty of trying to understand what these effects are becomes clear when one notes that the individual in modern society occupies many roles (family member, worker, church member, political actor) and numerous social positions (social class, race, urban or suburban residence). Not only is the white middle-class suburban male likely to be affected differently by technology than is the black lower-class urban female, but the categories themselves subsume an increasing diversity. Knowing that someone is a "blue-collar worker" or a "professional," for example, probably tells us less about the person than it once did. To be able to predict something about his behavior or attitudes as a worker, one would have to know more about the nature of the work he does, the size of the organization he works in, whether he has experienced much occupational or geographical mobility, and so forth. This is in itself a commentary on advanced technological societies.

It is easier to chart technological changes than it is to trace their sociocultural meanings. Perhaps it is this factor, in combination with the seemingly all-pervasive quality of modern technology, that has led to the popularity of the view that technology imposes its own nature and logic on society. In this regard it is well to remember some caveats offered by Robert Redfield in his classic work on *The Primitive World and Its Transformations.* Redfield distinguishes between the "moral order" and the "technical order." In precivilized or folk societies, the moral order was dominant. These societies were "held together essentially by common understandings as to the ultimate nature and purpose of life," by judgments about what is right. By contrast, "the bonds that coordinate the activities of men in the technical order do not rest on convictions as to the good life; they are not characterized by a foundation in human sentiments; they can exist even without the knowledge of those bound together that they are bound together. . . . In the technical order men . . . are organized by necessity or expediency." [1] One cannot, however, simply "declare that in civilizations the technical order predominates over the moral. . . . In civilizations the relations between the two orders are varying and complex." The changes in "the intellectual and moral habits of man . . . do not reveal themselves in events as visible and particular as do material inventions. . . . In folk society the moral rules bend, but men cannot make them afresh. In civilization the old moral orders suffer, but new states of mind are developed by which the moral order is, to some significant degree, taken in charge. The story of the moral order is attainment of some autonomy through much adversity." [2]

Perhaps, then, one may characterize our era as one in which "new states of mind" are being developed "through much adversity." It is a time in which the established categories of discourse are themselves being called into question. While some decry the decline of freedom in a technological society and others argue that modern man suffers from too much freedom, and while some bemoan the difficulties of forming a stable identity and others proclaim that man's identity is all too readily established in a society that is free of alienation, there remain still others who declare that "identity" is a myth and that traditional notions of "freedom" are becoming outmoded in the modern world.

The simultaneous increase of individuation and interdependence, the cognitive orientation, and the frequent change and mobility that pose problems for our polity and our values also affect the lives of individuals and raise questions about the relationship between individuals and society. The options and constraints associated with advanced technology likewise affect the individual, however indirectly.

PROBLEMS OF FREEDOM, IDENTITY, AND SOCIAL INTEGRATION

In what ways, then, might technology be responsible for the psychological changes and challenges facing modern man? The various answers that are offered are best evoked by letting those who offer them speak for themselves. In most general terms, Ferkiss has provided the following answer: "[A]bsolute power over himself and his environment puts man in a radically new moral position." Throughout history, concepts of freedom and identity have been rooted in the fact that the society, economy, and physical environment were difficult to alter. Because this is no longer the case, "in the era of absolute technology, freedom and identity must take on new meanings or become meaningless. Other men can change your society, your economy and your physical environment. . . . They can alter your identity by controlling how you are brought up and what your experiences are; they can even program your children genetically in advance of birth. But perhaps more disturbing is the fact that you can do all these things yourself: you can change your appearance or even your sex, your moods and your memories, you can even decide what you want your children to look like. But if you can be whatever you want to be, how will you distinguish the 'real' you from the chosen? Who is it that is doing the choosing? Not all of these changes affecting the nature of freedom and identity are yet practicable, but . . . the fact that postmodern man knows that they are coming must already begin to alter his self-image and his actions." [3]

In this view, it is unlimited possibility and an excess of freedom that pose a challenge for identity. If almost anything is or will be technologically feasible, choice becomes difficult to the point of creating a serious psychological problem. Keniston,

Slater, and Toffler share this perspective, although they select somewhat different aspects of contemporary society as being principally responsible. For Toffler, it is the constancy and rapidity of change, for Slater the extent of mobility, and for Keniston both change and the social fragmentation of technological society.

Toffler diagnoses man's condition today as one of "future shock": the physical and psychological distress that "arises from an overload of the human organism's physical adaptive systems and its decision-making processes." [4] Advanced technological society generates this problem because it is characterized by an overabundance of goods and ideas ("overchoice"), constant mobility, and "transience" (a rapid rate of "making and breaking relationships with the things, places, people, organizations and informational structures that comprise our environment"). In this type of society, Toffler argues, men "enjoy greater opportunities for self-realization than any previous group in history." [5] Yet this very freedom creates problems for social and psychological integration; "the multiplication of life styles challenges our ability to hold the very self together." [6]

According to Slater, mobility is the principal cause of problems of integration. He foresees the development of "temporary systems" in which the individual becomes detached from "enduring and significant relationships." Increased mobility will bring about an intensification of this process of "individuation"—"the separation of the individual from those permanent groups that provide him with ready-made values and traits and from which he derives his identity." [7] As a result of his loss of a sense of being embedded in a more or less permanent group, the individual will become "a part in search of a whole, feeling neither enough like others to avoid a sense of being alone and lost, nor sufficiently included in a stable pattern of differentiation to have a sense of himself as a distinguishable entity embedded in a pattern of other such entities. In a society that places a value on individualism, this inability to experience oneself leads paradoxically to a cry for *more* uniqueness, more eccentricity, more individuation, thus exacerbating the symptoms." [8]

In Keniston's assessment, identity formation is especially difficult today because the rapidity of change and the high value placed on it produce a lack of connection between "social history and individual history." Confusion and lack of identity result. If the skills, assumptions, and values of the past are no longer relevant, then the identifications that do emerge are "cautious, selective, partial, and incomplete." Moreover, in a complex society, there are no "package deals" that could relate the distinct roles into a unity. Hence, "increasingly, we must achieve—not discover—our identities, and create—not find—our homes." But achieving identity is difficult, because the social fragmentation of technological society "pushes toward psychic fragmentation: without institutions or ideals to support psychic wholeness, inner division is a continual danger. Indeed, if an American persists in seeking a life where inner wholeness is reflected in the outer consistency of his daily activities, he is almost inevitably led to repudiate the life led by most Americans, with its careful dissociation of work, cognition, and public life from family, feeling, and fun." [9]

The tolls that change, mobility, and differentiation take on the human psyche—as portrayed by Toffler, Slater, and Keniston respectively—have been the subject of commentary and analysis since the days of the Industrial Revolution. Because advanced technology exacerbates these phenomena, problems of psychic unity or identity appear to become more serious as well. In the past, individuals were rooted in well-defined social classes and roles. Today, social classes are weak and ill-defined because of the high degree of social differentiation, and because increased mobility further removes the individual from lasting social ties. Under such circumstances, the difficulties of achieving identity and integration into society are heightened. Some of the malaise experienced by Americans today may be traceable to the attempts at self-definition in apposition to others—the quest for uniqueness of identity—or conversely, to the attempt to find "community" or "rootedness" in relationships with others.

Whatever the frustrations experienced in such quests, however, the expectations of individuals about their individual rights and their claims to personal satisfaction appear to have heightened. As Hacker has noted, "The fact is that the egos of 200 million Americans have expanded to dimensions never before considered appropriate for ordinary citizens. More aggravating than the crowding that comes with sheer growth of population is the exacerbated sensation of congestion arising when the individuals who rub against each other have heightened evaluations of their own merit and keener sensitivity to such abrasions." [10] Expressions of such heightened expectations take the form of demands for more satisfying and meaningful work and for more economic and political power and participation.

The extent of personal dissatisfaction stemming from "psychic fragmentation," "future shock," or "temporariness" is difficult to gauge. The analysis of these problems often conveys an exaggerated sense of their severity. An equally, if not more, exaggerated tone seems to pervade the writings of those who take the opposite perspective and see in technological society the curtailment rather than the extension of individual freedom. Representative here are Marcuse, Ellul, Fromm, and Roszak.

For Marcuse, the problem in the relationship between the individual and society is not the difficulty of establishing an identity or finding meaningful integration into society, but rather the too-complete integration of man into his society. Advanced technological society, he argues, "sells" itself to the people, so that they "recognize themselves in their commodities" and "identify themselves with the existence which is imposed upon them and have in it their own development and satisfaction." [11] In short, "a comfortable, smooth, reasonable, democratic unfreedom prevails." [12] Instead of allowing "the technological processes of mechanization and standardization" to "release individual energy into a yet uncharted realm of freedom beyond necessity," our society operates "through the manipulation of needs by vested interests." [13] Thus, while Keniston sees the technological society as providing few objects worthy of commitment and therefore as generating alienation, Marcuse argues that in this society the very concept of alienation becomes questionable, since the individual has taken as his own the needs that have been imposed upon him by the system. The society is "one-dimensional" in both thought and behavior. Those "ideas, aspirations, and objectives that, by their content, transcend the established universe of discourse and action are either repelled or reduced to terms of this universe." [14] This is accomplished through the translation of any transcendent or oppositional elements into purely operational matters. Even "the works of alienation are themselves incorporated into this society and circulate as part and parcel of the equipment which adorns and psychoanalyzes the prevailing state of affairs." [15]

Ellul similarly sees our society as characterized by "unfreedom" and the incorporation of oppositional elements. However, he sees "technique" itself rather than "vested interests" as being responsible. "Technique" is defined as "the totality of methods rationally arrived at and having absolute efficiency (for a given stage of development) in every field of human activity." [16] Whereas in the past "technique was applied only in certain narrow, limited areas," today it is dominant.[17] No one technique or group of techniques is responsible for man's enslavement. Rather, it is the "convergence on man of a plurality, not of techniques, but of systems or complexes of techniques. The result is an operational totalitarianism; no longer is any part of man free and independent of these techniques." [18] Technique has this effect on man because it is automatic (it selects the means to be employed by its own laws), self-augmenting (its progress is irreversible), monistic (it combines to form a whole, no element of which can be detached from the others), universal (it produces the same results in every culture), and autonomous (it "tolerates no judgment from without and accepts no limitation").

Technique is not "neutral," because it brings with it its own "method of being used." Controlling technique is not possible, because even though it may be possible after a time to "separate the good results of a technical operation from the bad, . . . in the same framework, the new technical advance will in its turn produce further secondary and unpredictable effects which are no less disastrous than the preceding ones (although they will be of another kind)." [19] Thus, in the face of the human problems that technique has brought—"superhuman demands" for disciplined labor, modification of the human environment, creation of a mass society—"human techniques" have arisen to restore the harmony between man and his universe. Such "human techniques" as education, propaganda, "human relations," sport, amusement, and medicine serve to assure man's "integration into the body social . . . with the least possible friction." Human relations in industry, for example, were "developed to adapt the individual to the technical milieu, to force him to accept his slavery, to make him find happiness by the 'normalization' of his relations with his group. . . ." Unlike Marcuse, Ellul argues that "none of this arises from human malevolence, or from some 'system,' but from the simple fact that other techniques are sought to answer the problems of industrial mechanization." [20]

Fromm's conception of the way in which technological society reduces man's freedom and corrupts his identity combines elements from the perspectives of both Marcuse and Ellul. He maintains that the operation of the present technological system is programmed by two guiding principles: that technological "can" becomes technological "ought" and that maximal efficiency and output must be secured. Hence, both the autonomy of technique and the nature of the system are implicated in the resulting "human consequences." These are to "reduce man to an appendage of the machine, ruled by its very rhythm and demands" and to "transform him into *Homo consumens*, the total consumer, whose only aim is to have more and to use more." [21] Modern man is pathologically passive and suffers from a dangerous split between intellect and feeling.[22]

The split between the cognitive and the emotional, and the dominance of the "technocratic" mentality, form the subject of Roszak's attack on modern society. In a manner akin to that of Marcuse, Roszak argues that "politics, education, leisure, entertainment, culture as a whole, the unconscious drives, and even . . . protest against the technocracy itself: all these become the subjects of purely technical scrutiny and of purely technical manipulation." [23] But while Marcuse sees men as becoming slaves to such one-dimensional thought principally because of the economic needs that have been inculcated in them, Roszak sees their weddedness to the scientific world view as the central cause. "The distinctive feature," he says, "of the regime of experts lies in the fact that, while possessing ample power to coerce, it prefers to charm conformity from us by exploiting our deep-seated commitment to the scientific world-view. . . ." [24] The technocracy holds its peculiar power over men by means of "the myth of objective consciousness." This "myth," whose influence is so pervasive that we accept it without question, holds that only by "cultivat[ing] a state of consciousness cleansed of all subjective distortion" can one "gain access to reality." [25] Objectivity has become "the one most authoritative way of regarding the self, others, and the whole of our enveloping reality. Even if it is not, indeed, possible to be objective, it *is* possible so to shape the personality that it will feel and act *as if* one were an objective observer and to treat everything that experience presents to the person in accordance with what objectivity would seem to demand." [26] Under the auspices of objective consciousness, "we subordinate nature to our command only by estranging ourselves from more and more of what we experience, until the reality about which objectivity tells us so much finally becomes a universe of congealed alienation." [27]

This extended summary of representative perspectives on the fate of man's freedom and identity in technological society seems to indicate that, whatever their disagreements about the causes and nature of the problems, most commentators present a rather gloomy picture. It would appear that a more sanguine account of the effects of technology on the individual is currently out of fashion and would be likely to subject the writer to charges of "naive optimism." To be sure, many writers do take note of the fact that technology has brought the individual much good in the form of improved living conditions. But this concession is usually relegated either to a passing sentence of no consequence to the central argument or to a prelude to the declaration that such material progress has come at the expense of man's soul. It may well be that this trend in the literature is a healthy corrective to an earlier overoptimism concerning technology and its effects, since the recognition that there *are* problems in man's relationship to technology should not be ignored. Yet, insofar as much of this literature is equally one-sided, some more judicious assessments would seem to be needed.

For those who consider that little short of revolution can be efficacious in dealing with such problems, perspectives such as those of Ellul, Roszak, Fromm, and Marcuse may be the only ones that make sense. For those who are not so minded, these authors offer some cogent analyses and some useful insights, which, however, need to be tempered by other facts and interpretations.

Thus, it is difficult to accept the idea set forth by both Ellul and Fromm that technology is autonomous. As has been pointed out numerous times, technology is "wholly incapable of setting its own rules on the basis of its own logic within a completely closed circle. . . . Nor does every technological achievement carry within itself the imperative toward a further step. Mankind has selected an array of technologies that currently form its technological order, but has ignored others which continue to lie dormant. Undoubtedly we will drop some technologies now in use, and thus change still further the existing technological order, which in fact is necessarily in flux as it responds to the changing needs of mankind. . . ." [28]

The passivity of individuals and their acceptance of the social and technological status quo that both Marcuse and Fromm are concerned about seem also to be exaggerated, and are belied in part by current displays of activism. Nor does the dominance of objectivity, which pose some problems for values, mean that the subjective aspects of existence are overpowered or obliterated. It is not clear, for example, whether the family and other social structures devoted to the maintenance and support of the personality are being strengthened or weakened in modern society.

Keniston has noted that one of the reasons that adolescents today have difficulty in forming a clear identity or sense of self is that older models, such as those presented by their parents, are seen as being irrelevant. In the same vein, various commentators have observed that the family is in a state of decline. Thus, Morison argues that the family as an institution is suffering a major decline in both prestige and importance, because it is not a good transmitter of new knowledge, because its socializing and educational function is increasingly being assumed by the state, and because it is being weakened by concerns about population growth and the new knowledge of human genetics.[29]

Slater does not agree. He maintains that the "democratic family," which, for historical reasons, has long prevailed in the United States, is one of the mainstays of an advanced technological society. In fact, he predicts that all modern societies will eventually develop a similar family structure, since it is well suited to the needs of such societies. Like Morison, he notes that modern parents are "obsolete in a way that rarely befell parents of any previous century." But it is this very fact that makes the democratic family so important. Because this family rests on the assumption that "children may adapt better to their environment than did their parents, and that therefore their parents cannot take for granted the superiority of their own knowledge, perceptions, attitudes, and skills," it helps to foster adaptability to change.[30] Slater also contends that the importance of the marital relationship is growing. Because "the social ties of modern Americans are becoming so ephemeral that a permanent point of reference seems essential," it is likely that the marital relationship will be intensified.[31]

According to Sennett, the family, in attempting to provide the cohesion that is absent in the society at large, becomes a microcosm of the social universe and helps to shut its members off from participation in the larger society.[32] He condemns modern family living for channeling men's energies away from the public sphere, and hence supporting a pattern of public apathy that is fostered by a strong and distant government and the sway of technical expertise. It is probably true that the subjective reality for many individuals is one of inability to influence government decision—either because "they" will not listen or because "I" am incompetent to form intelligent opinions and policy suggestions.

Yet the demands for power and participation have, if anything, increased rather than decreased.

SOME PROPOSED SOLUTIONS

If one were to accept "Ellul's demon," man has virtually no capacity to alter or control his technology. Yet it is possible to view modern technology as a complex system without at the same time believing it to be beyond man's control. Thus, somewhat like Ellul, Toffler comments that "our problem is no longer the innovation, but the chain of innovations, not the supersonic transport, or the breeder reactor, or the ground effect machine, but entire inter-linked sequences of such innovations." [33] But he sees a possible solution in the attempt to anticipate the secondary consequences of technological change. While admitting that "we can never know all the effects of any action, technological or otherwise," he argues that we are not helpless, since "it is, for example, sometimes possible to test new technology in limited areas, among limited groups, studying its secondary impacts before releasing it for diffusion." [34] It is this idea that underlies the various proposals to establish mechanisms of technology assessment.

A related idea is that proposed by Mishan and others. Mishan contends that the policy of economic growth and "the indiscriminate pursuit of technological progress" have been responsible for "destroy[ing] the chief ingredients that contribute to men's well-being." [35] Therefore, a reversal of these policies is required; "more selective criteria of welfare" should be developed. Legislation should be established that would recognize the individual's right to such amenities as privacy, quiet, and clean air, and resources should be diverted from "industrial gadgetry" to "re-planning our towns and cities" and "re-creating" the environment.[36]

Fromm's "revolution of hope" consists in the replacement of alienated bureaucracies with humanistic management that would allow for grass-roots activity, and in the development of a system of "humanistic planning." Goodman's "revolution" is of a different sort. Science and technology, he argues, are not intrinsically harmful. Rather, "they have fallen willingly under the dominion of money and power. Like Christianity or communism, the scientific way of life has never been tried." [37] What is needed is more prudence in the development and use of technology, concern for the ecological fit of technology, and decentralization of the decision-making processes associated with technological invention and use. Such changes will require "a kind of religious transformation. Yet there is nothing untraditional in what I have proposed; prudence, ecology, and decentralization are indeed the high tradition of science and technology. Thus the closest analogy I can think of is the Protestant Reformation, a change of moral allegiance, liberation from the Whore of Babylon, return to the pure faith." [38]

Other solutions concentrate less on technology and more on man and his culture. For Roszak, the answer lies in the rejection of our scientific and technological culture and the development of a "counter culture." In place of objective consciousness, the counterculture urges the development of a new culture based on the "non-intellective capacities of the personality," on visionary experiences. It pits the "immediacy of the personal vision" against the "aloofness of objective knowledge," [39] the magical against the scientific, and argues that the two cannot really coexist, for the "schizoid" attempt to maintain a little of both allows us "to throw off flurries of intellectual sparks, but short-circuits any deeper level of the personality." [40] The counterculture thus celebrates a kind of "truth" that cannot be obtained through rationality. It is a subjective, and therefore highly personal, truth. Their solution, then, to the problem of alienation is not to make social participation more meaningful, but rather to make it irrelevant by substituting a purely personal existence.

It is hard to see how the problems of an expert-ridden, quantitatively oriented society are to be solved through this kind of mystical withdrawal. If technological consciousness is a psychological and social burden, if it presents a challenge to the maintenance of meaningful social participation and ultimate values, then attempts must be made to build new meanings and values into social life.

While Roszak does not attempt to reconcile the mystical withdrawal of the counterculture with the need for social change, Reich does.[41] Reich's argument is that when the emerging new consciousness gains sufficient strength, it will bring about a collapse of the current social system. Advanced industrial society, he maintains, requires a bureaucratic, technocratic mentality to sustain it. Hence, as soon as a sufficient number of people reject this mentality, the system will be brought to a halt. In fact, the revolution has already begun, in part because of the contradiction between the demands on workers to adhere to a strict Protestant Ethic and the demands on consumers to be hedonistic, and in part because the affluence of advanced modern society allows for the possibility of a new way to live—for equality, love, and an end to the antagonisms generated by capitalistic competition. No attention is paid in Reich's schema to what the social structure will look like after the revolution takes place or to how the movement will accomplish its ends without a fight.

While Slater does not reject the existing culture, his emergent "temporary systems" also stress feelings and spontaneity. Since groups will form and dissolve rapidly, he argues, "it will be increasingly necessary to take people as one finds them—to relate immediately, intensely, and without traditional social props, rituals, and distancing mechanisms. . . . It seems clear that one of the unintended functions of 'sensitivity training' or 'basic encounter' groups is anticipating a world of temporary systems, since these groups emphasize openness, feedback, immediacy, communication at a feeling level, the here-and-now, more awareness of and ability to express deeper feelings, and so on." [42]

If problems of identity bedevil many contemporary observers, at least one analyst has come to believe that "identity" itself may be a myth. Bennett Berger argues that the notion of a "search for identity" helps "accommodate [adolescents] to their juvenile status and therefore constrains them to keep out of markets and areas where only the 'mature' belong." This has become necessary in industrial societies as the need for the relatively unskilled labor of the young has declined. Although the "identity myth" proclaims that at some point one "finds" one's identity, this is but a social convenience. "There is no self-evident psychological reason for the stormy search to end at any point before the grave, since presumably personal growth and change continue as one goes on accumulating significant experience." Now, "if identity is inherently something elusive or ungraspable, if the self is an onion rather than a nut, then the most important function performed by institutionalizing the search for something unfindable is to induce the sort of anxiety which promotes the mobility of the psyche. A person who does not know who he is might just be anything, and hence is fit for the unanticipatable opportunities and eventualities which rapidly changing industrial societies provide. . . . Is it too radical to entertain the idea that we may be moving into a period when men and women will be called upon to sustain greater and greater disjunctions among and between the different dimensions of their lives, and that to urge the development of an integrated identity is to exacerbate rather than alleviate the problem? Isn't the 'individual' whom many of us admire most precisely the person who sustains the most unusual combination of social statuses and hence who lends to each of his role performances some unexpected features derived from his other lives—in short, the man who never finds *an* identity because he is too busy accumulating being. . . ." [43]

Just as Berger proclaims identity to be a myth, Lifton detects the emergence of a new type of man who is characterized precisely by the absence of any stable identity. "Protean man" exhibits a style of life "characterized by an interminable series of experiments and explorations . . . each of which may be readily abandoned in favor of still new psychological quests." [44] He is attracted to change and newness, while at the same time he is "drawn to an image of a mythical past of perfect harmony and prescientific wholeness." [45] The causes of this personality structure may be traced to the breakdown of traditional cultural symbols and "the flooding of imagery produced by the extraordinary flow of post-modern cultural influence over mass communication networks." [46] Although ridden with ambivalences, protean man may turn out to be a relatively stable personality type that can deal effectively with the modern world.

□ □ □

1. Robert Redfield, *The Primitive World and Its Transformations* (Ithaca, N.Y.: Great Seal Books, 1953), p. 21.
2. *Ibid.*, pp. 24–25.
3. Victor C. Ferkiss, *Technological Man: The Myth and the Reality* (New York: George Braziller, Inc., 1969), pp. 21–22.
4. Alvin Toffler, *Future Shock* (New York: Random House, Inc., 1970), p. 290.
5. *Ibid.*, p. 283.
6. *Ibid.*, p. 284.
7. Philip E. Slater, "Some Social Consequences of Temporary Systems," in Warren G. Bennis and Philip E. Slater, eds., *The Temporary Society* (New York: Harper & Row, Publishers, 1968), p. 79.
8. *Ibid.*, p. 81.
9. Kenneth Keniston, *The Uncommitted: Alienated Youth in American Society* (New York: Harcourt Brace Jovanovich, Inc., 1965), p. 269.
10. Andrew Hacker, *The End of the American Era* (New York: Atheneum Publishers, 1970), p. 31.
11. Herbert Marcuse, *One-Dimensional Man* (Boston: Beacon Press, 1964), pp. 9 and 11.
12. *Ibid.*, p. 1.
13. *Ibid.*, pp. 2 and 3.
14. *Ibid.*, p. 12.
15. *Ibid.*, p. 64.
16. Jacques Ellul, *The Technological Society* (New York: Alfred A. Knopf, Inc., 1967), p. xxv.
17. *Ibid.*, p. 64.

18. *Ibid.*, p. 391.
19. *Ibid.*, p. 107.
20. *Ibid.*, pp. 355–56.
21. Erich Fromm, *The Revolution of Hope: Toward a Humanized Technology* (New York: Bantam Books, Inc., 1968), pp. 39–40.
22. *Ibid.*, pp. 41–42.
23. Theodore Roszak, *The Making of a Counter Culture: Reflections on the Technocratic Society and Its Youthful Opposition* (New York: Doubleday & Company, Inc., 1969), p. 6.
24. *Ibid.*, p. 9.
25. *Ibid.*, p. 208.
26. *Ibid.*, p. 216.
27. *Ibid.*, pp. 232–33.
28. R. J. Forbes, *The Conquest of Nature: Technology and Its Consequences* (New York: Praeger Publishers, Inc., 1968), p. 73.
29. Robert S. Morison, "Where Is Biology Taking Us?" *Science*, 155 (January 27, 1967), 429–33.
30. Philip E. Slater, "Social Change and the Democratic Family," in Bennis and Slater, *The Temporary Society*, p. 22.
31. Slater, "Some Social Consequences of Temporary Systems," *op. cit.*, p. 90.
32. Richard Sennett, "The Brutality of Modern Families," *Trans-Action*, 7 (September 1970), 29–37.
33. Toffler, *Future Shock*, p. 389.
34. *Ibid.*, p. 388.
35. E. J. Mishan, *Technological Growth: The Price We Pay* (New York: Praeger Publishers, Inc., 1970), p. 107.
36. *Ibid.*, p. 163.
37. Paul Goodman, "Can Technology Be Humane?" *New York Review of Books*, 12 (November 20, 1969), 33.
38. *Ibid.*

39. Roszak, *The Making of a Counter Culture*, p. 264.
40. *Ibid.*, p. 256.
41. See Charles Reich, *The Greening of America* (New York: Random House, Inc., 1970).
42. Slater, "Some Social Consequences of Temporary Systems," *op. cit.*, p. 86.
43. Bennett M. Berger, "The Identity Myth," lecture delivered at Forest Hospital, Des Plaines, Illinois, January 1968, mimeographed, pp. 11–13.
44. Robert Jay Lifton, "Protean Man," *Partisan Review*, 35 (Winter 1968), 17.
45. *Ibid.*, p. 25.
46. *Ibid.*, p. 16.

TECHNOLOGY and SOCIAL PROBLEMS

In examining the effects of technological change on American politics, values, and individual psychology, the preceding three chapters have touched upon some of the most fundamental social problems of our time: the governance of a complex interdependent society and the attendant difficulties of providing an adequate supply and distribution of public goods and of balancing democracy against expertise; the conflicts of values and the stresses of value change; the problems associated with individualism, freedom, and social integration. These broad issues provide the framework within which to understand the more narrowly defined social problems. The problems of medical care, of work and leisure, and of the city discussed in this chapter reflect these underlying issues and the associated need for a better adjustment among technologies, social structures, and values. In each of these areas, there is need for reorganization and redefinition—for changing the concept and nature of urban government, the organization and delivery of medical care, the mechanisms of labor-force supply and demand. Questions are being raised about what a city should be, about the meaning of work and leisure, and even about the definition of death. The problems of planning and of value tensions discussed earlier are amply illustrated by these concrete cases.

Medical Care

The major issues that result from advances in biomedical science and technology may be seen as a microcosm of the kinds of problems found in the more general technology and society area. If biomedical problems are unique, it is because they impinge more directly and evidently on the physiology and psychology of man than do other sciences and technologies. Because of this closeness to man, the social and ethical dilemmas posed by biomedical science often appear particularly visible and poignant. The direct relationship between biomedical science and health also poses special problems for science policy—that is, for organization and the allocation of resources. In contrast to research in nuclear energy, for example, where the ramifications range over a variety of industrial and military areas, the implications of biomedical research have a direct and almost exclusive bearing on health and medical concerns. The relationship between science and extra-scientific considerations is therefore more direct here than in other areas of science policy. Disregarding such special characteristics, however, one can discern four major themes of relevance in the biomedical sphere, as well as more generally in the technology–society relationship: (1) the shift from private to public responsibility, (2) the "crisis" in the social service sector, (3) the response of the legal structure, and (4) problems of technology assessment.

THE SHIFT FROM PRIVATE TO PUBLIC RESPONSIBILITY

It is now a commonplace that good medical care, once considered a privilege, is coming to be considered a right of every citizen. In this transformation, the simple doctor–patient relationship has been superseded by a tripartite relationship, in which the society, or the state, bears ultimate responsibility for the care of the patient. Such intervention manifests itself at many different levels: in the establishment of public medical facilities, in the provision of health insurance, in the support of medical education, and in the financing of biomedical research. Through the power of the purse strings, the government can and does exercise control over the nature and direction of medical care. Medicare legislation authorizes the government to stipulate what kinds of medical services will be paid for. It thereby grants the government some measure of quality controls and allows it to impose standards in matters of cost and organizational structure. Through support of medical education, the government may encourage the development of certain studies, in public health and community medicine, for example. By its substantial involvement in biomedical research, moreover, the government influences the direction of scientific endeavors. As in other spheres, increased government involvement in medical R&D is necessary because the risks of development are large and the rewards are unattractive to private capital. "At the present time, the Federal Government provides about 60% of all funds spent in the U.S. for health research. Industry contributes about 28% and foundations, voluntary agencies and others contribute about 11%." [1]

Government involvement raises the question of the optimal interaction of public and private institutions in financing and organization. As with all public or quasi-public goods, the ideal is to institute public responsibility while maintaining the flexibilities and competitive efficiency afforded by private enterprise.

A second series of questions posed by the assumption of public responsibility for health is concerned with the trade-offs and balances between individual and social benefits. As it becomes increasingly legitimate for the government to legislate for public health, conflicts between the rights of the individual and the good of society increasingly arise. Some of the controversy surrounding the relationship between smoking and health may be taken as an example. It is not clear how far the government can or should go in imposing restrictions on individual behavior, particularly when that behavior has only indirect effects on the welfare of others (for example, the economic problems caused by the early death of a breadwinner or the bad example set for one's children). Controversies surrounding what is or is not legitimate in experimentation with human subjects likewise involve weighing the individual good against the social good. Some commentators have suggested that codes of ethics for the regulation of human experimentation should take explicit account of such considerations by imposing more stringent constraints on experiments that are likely to have small payoffs than on those that might result in significant scientific breakthroughs. Other writers assert the primacy of the rights of the individual and argue that scientific progress, whatever its ultimate benefits to mankind, is not the ultimate good, so that the rights of individual subjects must be carefully protected no matter what the costs to science may be.

In the health sphere, as elsewhere, the assumption of public responsibility has called forth the need for more deliberate and conscious choices with respect to the allocation of resources. The choices are of several sorts: How much should we allocate to medical care as opposed to research? On which diseases should research efforts be concentrated? Within the resources devoted to medical care, what is the proper distribution of benefits among various population groups?

The high value placed on health and the focus on disease-related research has led researchers to fear that an increasingly larger share of the national health effort will be devoted to medical care and to applied research at the expense of basic research. They view this prospect as particularly dangerous because much of biomedical research remains in what Alvin Weinberg has called the "prefeasibility stage"; that is, the feasibility of application has not been demonstrated, and therefore the link between "basic" and "applied" research is not as clear as in the physical sciences.[2]

The major factors influencing the allocation of resources to research and development are the scientific state of the art in each specialty and the social importance of the problems under investigation. These two criteria are often in tension, since difficult trade-offs must be made between the desire to advance scientific knowledge in a particular field and the desire for direct social benefit. It is no easy matter to decide whether to support research that promises a fast therapeutic payoff or to emphasize research in areas where advances may be slower but also more significant in the longer run. In addition, research efforts might be directed toward the cure of diseases that afflict the old or of those that afflict the young. Dilemmas of this kind compound the usual difficulties involved in the allocation of research money.

The dilemmas of resource allocation in medical care are illustrated most starkly by figures such as the following: The estimated cost for transplanting a kidney and providing follow-up care is $5,000, while $130 per year could give adequate routine health care to a poor American; $1 billion could buy enough kidney dialysis centers to serve the 25,000 or more people who will need them in the next decade, or it could buy comprehensive ambulatory health care for more than 1,250,000 poor people.[3] Thorny problems arise too in the selection of patients to receive such rare, expensive, and life-sustaining treatments as dialysis.

THE "CRISIS" IN THE SOCIAL-SERVICE SECTOR

While the specific underlying causes of the "crisis in medicine" may differ from those responsible for such other "crises" as those in education or urban affairs, there are some important similarities. In each of these areas, the "crisis" stems primarily from the rising expectations and increased demand for welfare services in an affluent technological society. Although "supply" may increase to meet increased demand, the quality of the services in question does not keep pace with the expectations, and their distribution remains inequitable.

Thus, although health is the third largest industry in the nation, and although annual health expenditures comprise approximately six percent of the GNP, the health care received by the poor remains inadequate (as measured, for example, by infant mortality rates and ambulatory care facilities). Improvements in the quality of medical care have clearly been made, but these are often accompanied by greatly rising costs, inadequate geographical distribution, and the increased difficulty of access to such services as those of a private physician. Since 1961, over one-third of the increase in personal expenditures for health has been due to the higher cost of medical care.[4] The increased specialization and assumption of administrative and managerial responsibilities among physicians have resulted in a net decrease of the number of physicians available to provide family care. The disproportionate concentration of health services in well-to-do urban areas has left rural dwellers and the urban poor inadequately serviced.

The problem of rising costs is not amenable to the type of solution so often applied successfully by industry—that is, the introduction of new technology. For in health, as in some other service sectors, technology does not cut labor costs. To the contrary, new medical technologies often result in the need for more and higher-paid staff, over and above the expense of the equipment itself. Rising labor costs have also contributed to the costs of medical care as the traditionally underpaid hospital workers have been gaining greater bargaining strength. (This trend is paralleled in the field of education, where most of the technology employed is not labor-saving and teachers' salaries have been increasing as a result of unionization.)

The crisis in medicine might thus be seen as primarily a crisis in organization. The problem of costs might be attacked through seeking greater efficiency, through elimination of wasteful duplications of facilities, or through economies of scale that could be effected by greater coordination of services. Much of the burden of hospital costs is currently assumed by third-party payers who exert little or no control over prices, so that there are few incentives for economy and efficiency.

Whatever the shortcomings of the application of cost-accounting techniques in the Department of Defense, that institution could at least muster some degree of agreement on goals and did not have to contend with the repugnance "on principle" to the use of those techniques. By contrast, the very notion of applying cost-accounting criteria to health matters is anathema in some quarters and raises devilish questions about the comparative merits of alternative health goals. Moreover, the nature of the health market is such that no matter how inefficiently medical institutions may be run, they are unlikely to "fail" or be driven out by competition. As a result, the health sector is riddled with such anomalies as "two new hospitals, both half-empty, stand[ing] within blocks of each other" or "a half-dozen hospitals each equipped and staffed to do open heart surgery—one of the most expensive of all surgical specialties . . . [despite the fact that there are] barely enough cases to keep one of the centers busy."[5]

To overcome such gross inefficiencies, community and regional planning might be required. But such planning invariably faces not only the resistance of entrenched interests, but also the need for restructuring existing political and administrative jurisdictions. An effective reorganization of health-care facilities might proceed in accordance with health service areas rather than with established political boundaries. The "community of solution" for some health problems may cross state lines and perhaps even national lines, as in the cases of smallpox and malaria. Attempts to move the health care system in the direction of regional planning have received legislative encouragement, but have not yet produced much change.

The "crisis in medicine" would seem to require more than just a reorganization of the health-care system; a reorganization of the health profession might be needed as well. As advances in biomedical knowledge and technology have occurred, an increasingly complex system of medical specialization has arisen. To provide for the integration of these specialized services, several types of "generalists" are now needed. The much-discussed decline of the general practitioner may be taken as a first example. To some extent, this problem is currently finding a solution in the tendency of specialists in internal medicine to assume some of the functions of the vanishing family doctor. But given the great demand for and the unabated proliferation of specialists, another possible solution to this problem is gaining favor: the training of upgraded medical assistants. The role of this new type of health professional would be to deal with the "whole patient," thereby providing psychological comfort as well as serving as liaison between the patient and the appropriate specialists.

A second new type of generalist is the public health physician. As the focus of health efforts shifts from exclusive concern with the individual patient toward community health, there is a need for physicians able to set priorities in the allocation of health resources within the community and to cooperate with other segments of the community in order to secure the social and economic prerequisites for health. Such physicians would require training in the social sciences and in the techniques of statistics and cost accounting. Finally, as medicine comes more and more to rely on sophisticated technologies, there will be an increased need for physicians capable of dealing with the relation between engineering technology and medicine.

THE RESPONSE OF THE LEGAL STRUCTURE

Legislation specifically related to health and medicine has taken two major forms: policy setting for the financing and organization of medicine (for example, Medicare and the Comprehensive Health Planning Act) and regulation of the nature and practice of medicine (for example, mandatory vaccinations for smallpox). But the challenges facing the legal structure as a result of developments in the biomedical sciences are representative of problems emerging from the growth of science and technology more generally.

The growth and diffusion of science and technology present lawmakers with an ever-expanding body of highly complex scientific developments about which they are called upon to legislate. To cope with this challenge, a vast network of information and advisory bodies has been developed to counsel the Congress. But adequate information provides only part of the solution. Problems of communication between scientists and lawmakers often arise as a result of the differing outlooks and constraints of the two groups. The goals and interests of the congressman and the scientist may be convergent or contradictory depending on the specific issue involved, so that lawmakers, although relying on testimony from scientists, may and often do interpret this information to suit their own purposes and predilections. In the controversy over possible legislation on cigarette smoking, for example, many congressmen exaggerated and capitalized upon the disagreements among scientists, in order to support their disinclination to enact strong regulatory measures. By contrast, in the case of legislation calling for the diagnosis and treatment of phenylketonuria (PKU)—a rare disease that causes mental retardation in children—state legislators downgraded or disregarded existing disagreements among scientists. Scientists, in turn, may slant their testimony for purposes of demonstrating their success and appealing for more research money. To what extent communication failures between scientists and legislators are unavoidable and to what extent they result from deliberate distortion remains an open question. In either case, it would seem incumbent on both parties to assume greater responsibilities for communication and cooperation.

The case of the PKU legislation illustrates still another dilemma for the law. In attempting to protect the health and wellbeing of the citizenry, legislators face the danger of freezing the existing scientific state-of-the-art into law. It is characteristic of science to be continually changing, whereas the law tends to be conservative. Therefore, in matters pertaining directly to scientific knowledge, such as the treatment of a given disease, lawmakers need to tread a fine line between taking no action (where intervention may be desirable or necessary) and enacting legislation that is ultimately self-defeating (by requiring specific actions that science later discovers to be deleterious). The consensus here would appear to be in favor of risking sins of omission rather than of commission. Alternatively, it might be possible to include review mechanisms or cutoff dates in legislation of this type.

Yet another category of legal problems stemming from scientific and technological developments is illustrated by the cases of human experimentation and organ transplantation. Legislative and judicial policy on such matters has not yet been set. No laws have been devised to cope with the questions that have arisen in the wake of the great expansion of clinical and drug research. At present there exist only professional codes of ethics (primarily the Nuremberg Code and the Helsinki Declaration) and governmental guidelines (primarily those of the U.S. Surgeon General and the New York State Board of Regents). In starkest terms, the dilemma for the law may be stated thus: Can or should a system of ethics be codified into law? Whatever the perils of such a procedure might be, it is unreasonable to assume that the law can remain aloof from the issue. Some degree of legislative codification will become necessary as more and more cases are brought before the courts and as the need to establish a basis for judicial decisions becomes more pressing. The major problem involved in adhering to a set of general ethical guidelines is the difficulty of translating the general criteria into workable guides in specific clinical situations. Because of the great variety of experiments, experimental situations, and potential hazards and benefits, the same general precept may have quite different meanings and ramifications in different situations. For example, does "consent" really mean the same thing when the subject is under the psychological duress of a serious illness as it does in the case of a healthy subject? Does "informed consent" mean the same thing when outcomes can be predicted with a high degree of certainty as when they cannot? Moreover, the notion of balancing potential dangers against potential benefits that is found in both the Nuremberg Code and the Surgeon General's guidelines is an exceptionally difficult one to apply. A codification of guidelines into law would have to recognize both the moral sensitivities of the community at large and the necessity to balance the rights of subjects against the rights of the investigator. The constraints on experimenters should not be so great as to stifle scientific and medical progress, since there are often social and individual costs to be paid for failing to experiment. Because of these difficulties, it is probably well to be cautious about instituting legislation in cases where informal control mechanisms are operating effectively. Experimentation with various forms of informal control can provide the time needed to air the issues in their full complexity and, it is to be hoped, to "debug" the systems before they are set into law.

The problems of human experimentation are complicated by the fact that there is often a fine line between "experiment" and "therapy." As Renée Fox has expressed it, "all physicians trying out new procedures or drugs on sick persons . . . [face] the question of determining how experimental and/or how therapeutic an innovational treatment is, and in the light of that evaluation, deciding under what circumstances and on whom it may justifiably be used." [6] This dilemma has been particularly acute in the case of organ transplantation. Unlike the case of human experimentation, however, legislation of relevance to transplantation already exists—but it is premised on earlier technologies (more specifically, on the absence of the technology for effecting organ transplants). Many state laws concerning the disposition of the body of a deceased predate the advent of organ transplants. Moreover, although thirty-one states now allow prior organ donation, most physicians are reluctant to remove organs without specific permission from relatives of the deceased, because the laws have not been tested in the courts. Legal clarification of the rights of the deceased and his next of kin will no doubt be forthcoming. It might be noted in this regard that despite traditional notions of the sanctity of the body, a recent Gallup poll found that 70 percent of the persons questioned expressed a willingness to donate their bodies for medical purposes.[7]

Much thornier legal and ethical problems arise in connection with the definition of death. The development of artificial circulatory and respiratory devices has enabled us to keep people alive after natural physiological processes have ceased to function. This has raised many ethical problems: When should artificial maintenance of life be employed and for how long? When might it be appropriate for the benefit of all concerned to unplug the machines? Although it is not legal, euthanasia has often been tolerated when it could be effected by simple acts of omission. The increasing reliance on artificial devices, however, means that euthanasia could no longer be effected except through deliberate acts of commission.

Definition of the precise moment of death is central to the legal and ethical ramifications of transplantation. If the criteria for determining death remain unclear, then there are no guidelines for deciding when a physician may legitimately remove an organ for transplantation. Since deterioration of organs occurs even while artificially sustained "life" still persists, a concern for the benefit of the potential recipient would dictate the earliest possible removal of the organ. For this reason, some commentators have maintained that it is ethically legitimate to remove the needed organs before the life-sustaining machines have been turned off, once it is determined that the patient cannot be saved. The relevant question would then become one of determining when there is no longer any possibility of restoring the patient to life, rather than of defining the moment of death. Supporters of this view note that patients who have been kept alive by means of artificial devices appear upon autopsy to have been dead for some time. Opponents fear that the use of any definition of death that does not include the traditional cessation of circulation would cause the whole transplantation procedure to fall into disrepute. They argue that the question of when to stop the machines and let the patient die is a separate question from that of deciding when the patient is dead. Until death occurs, interference with the body is illicit. The law is thus faced with the dilemma of taking into account modern scientific knowledge and technology in redefining what constitutes death or of maintaining the current definition of death. Retention of the current definition of death would incur the risk of stifling further medical progress and of losing the lives of potential organ recipients. A new legal definition of death would require the institution of safeguards against unethical practices. The most prominent attempt to provide a new definition of death is that of the Ad Hoc Committee of the Harvard Medical School, initiated and directed by Dr. Henry Beecher. The report of this committee defined "irreversible coma as a new criterion for death" in persons "who have no discernible central nervous system activity." It also specified the "characteristics of a *permanently* nonfunctioning brain" that should be used to determine whether or not death has occurred.[8]

PROBLEMS OF TECHNOLOGY ASSESSMENT

The difficulties of applying some form of technology assessment are well illustrated by the case of biomedical technology. Not only are we frequently unable to predict the consequences of many new technologies, but there is the further and in some senses prior problem of a lack of clarity about values and goals. It is extraordinarily difficult to assign priorities, to weigh the various desiderata against each other. Clearly, the two problems—lack of knowledge and lack of clear values and goals—feed each other. If we knew what the effects of making no deliberate genetic interventions would be, for example, it might be somewhat easier to weigh the values of individual freedom and dignity against the values of human improvement. On the other hand, it is conceivable that such better knowledge of outcomes might serve to exacerbate value conflicts.

The value dilemmas involved may be characterized as (1) short-range versus long-range considerations, (2) balance of individual and social benefits, and (3) containment of the negative effects of science and technology versus freedom of research and inquiry. Any given scientific and technological development might be highly desirable currently, even though its long-term consequences (often only dimly seen) might be disastrous. For example, some possibilities of genetic change that might be desirable now might have the effect of "locking in" certain traits that could be highly problematic in the future.

The problem of unexpected and unintended consequences of technological development remains a major difficulty in any attempt to forecast and anticipate long-term effects. Few people, for example, foresaw that the obvious benefits resulting from better nutrition, sanitation, and health care would result in a vast growth of population that now threatens to outrun society's ability to deal with it. The longer-term consequences of some drugs or of organ transplantation remain problematical. What, for example, might be the social consequences of removing the technological, ethical, and legal constraints on transplantation and of establishing a system of organ banks? In the short term, the beneficial effects on the health of the nation could be great. But how is one to assess the possible long-term consequences of a technology that would allow for the indefinite preservation of life through the replacement of vital organs?

As noted above, the problem of balancing individual and social benefits is particularly difficult with respect to human experimentation. While small risks to the individual for the benefit of the society are likely to be tolerated, jeopardy to the person will probably not be, no matter how great the potential social benefit. Society has not yet begun to face this issue in such matters as population control. While the need to limit population growth is recognized as important, the idea of placing restrictions on the freedom of individuals to procreate is anathema to many people.

The dilemma of permitting freedom of research versus containing the negative effects of science and technology is well illustrated by the cases of genetic and behavior-control research. The idea that research in these areas could turn out to be disastrous has led some people to advocate what might be called a "stop technology" movement. Even some members of the Harvard Medical School team that succeeded in isolating the gene structure of a bacterium have proclaimed that they would not pursue that line of research if they had it to do over again.[9] The issue cannot be dismissed so simply, however. There are numerous methods that might be employed to control science and technology. But it is to the *uses* of such science and technology that control efforts must be devoted, not to the ability to do research and to engage in inquiry. Some control over the research process is of course required in order to prevent unethical practices, but the idea that some kinds of research should be forbidden entirely could lead ultimately to the very thought control that the critics of genetic and behavior-control research fear.

Technology assessment in the biomedical sphere often appears to trail off into questions of ultimates. A consciousness of this fact has led physicians and scientists to call for the assistance of social scientists, philosophers, and theologians. It has also been suggested that it is incumbent upon all concerned to educate the public in these matters, so that an informed citizenry may help to shape the decisions that will have to be made. The possibilities of deliberate construction of the human species through genetic engineering and of substantial prolongation of life through transplants and artificial organs have made recent science fiction seem less remote than it had been thought. It is to be hoped that our technical virtuosity will be matched by the degree of social wisdom needed to control its consequences.

Work and Leisure

The pace of technological change in industry has been a source of much public concern. In the late 1950's and early 1960's, the literature on the subject was often characterized by the extremes of grave pessimism concerning the masses of people who would be "thrown out of work by machines" and utopian optimism about the "leisure society" in which man would at last be freed from "the burdens of labor." By the mid- and late 1960's, more sober views began to prevail. In 1966 the National Commission on Technology, Automation, and Economic Progress issued a comprehensive report on the subject. It concluded that "technology eliminates jobs, not work." That is, "the general level of unemployment must be distinguished from the displacement of particular workers at particular times and places." [10]

The amount of unemployment is determined by the interactions among three factors: growth of the labor force, increases in productivity or output per man-hour, and growth of total demand for goods and services. As a result of technological changes that lead to productivity increases, either less labor is needed to produce a given amount of product or more goods and services can be produced with the same number of man-hours. Therefore, employment will decline unless there is a growth in output. Since the labor force grows too, however, the growth in output (gross national product) must be more rapid than the *sum* of productivity increase and labor-force growth in order for the employment level to be sufficient to absorb the growth in the labor force. Therefore, the Commission recommended "aggressive fiscal and monetary policies to stimulate growth" in order to maintain a high level of employment.[11]

Complacency about the effects of technological change on the level of employment is certainly not warranted, however. In addition to the difficulties involved in maintaining a high level of employment, problems of the displacement of specific workers and of mismatch between labor-force supply and demand remain. In addition, a renewed concern has emerged about the quality of work and leisure in an advanced technological society.

THE STRUCTURE OF THE LABOR FORCE

The significance of economic policy in determining the employment effects of technological change serves to illustrate the general importance of social structures and attitudes in mediating between technology and its effects. Such large-scale changes in the occupational structure as the growth of the service sector and the increased importance of education are also the result of an interaction between social and technological factors. Technological advances in the production and processing of goods have allowed the growing demand for goods to be fulfilled quite adequately by an increasingly smaller proportion of the total work force. At the same time, the introduction of advanced technology into the service sector has not resulted in a reduction of labor needs, so that there has been some growth in service-sector employment as the demand for labor elsewhere has decreased. But technology is only indirectly responsible for the growth in demands for services. This demand has grown as the society has become more affluent. Generally, as income has risen, the demand for services has increased more rapidly than the demand for goods. Moreover, growing affluence has supplied the basis for expanded efforts to provide a greater proportion of the population with such services as health and education.

The enhanced importance of education as a prerequisite to employment is likewise a product of both social and technical factors. In a society with sufficient wealth to allow large numbers of people to postpone entry into the labor force so that they may receive more education, employers often have a sizeable pool of educated manpower at their disposal. Given the value placed on education in our society and the ease with which educational credentials may serve as a screening device, employers will often prefer to hire the more educated, even though the intrinsic requirements of the job do not call for such education. It is certainly true that as technological advances have diminished the need for unskilled manual labor, the newer jobs tend to require more education and training. Education also tends to produce a greater flexibility in the worker, although there has been some recognition that "overqualified" manpower may create problems. There are many indications, however, that the educational requirements for a given job are often exaggerated.

James Bright has been so impressed with the interaction between social and technical factors in the work sphere that he argues that while automation affects the skill level required for certain jobs, the available pool of manpower skills in turn influences the progress of automation. In his "law of automation evolution," he asserts that machinery evolves to provide the degree of automatic operations that is "economically supportable by the level of skill that can be made readily available in the existing work force." Thus, "when the machine manning needs have been reduced to a standard that is normally available in the local work force, . . . the economic incentive for automation progress disappears." [12]

The progress toward a totally automated industrial system is in fact nowhere near as rapid as had been feared; "only a small proportion of the labor force is employed in industries at either the beginning or the advanced stages of automation." [13] Taking account of "those industries . . . which appear to be moving most rapidly toward automation—printing, rubber, insurance, machinery, textiles, primary metals, motor vehicles, aircraft and missiles, and food-processing industries such as grain mills and bakeries" only adds a small additional percentage to the proportion of the labor force directly affected by automation.[14] Yet the long-range effect of technological change on the occupational distribution of the labor force is clearly in the direction of the progressive elimination of the least-skilled occupations and the growth of professional, technical, and white-collar work. This basic change that technology has effected in the occupational structure is most often symbolized in the distinction between the pyramid-shaped structure of the early industrial labor force and the diamond-shaped structure of the modern industrial system. As the repetitive manual labor that constituted the base of the pyramid has been replaced by machine power, the base has shrunk, and there has been a widening near and below the top that reflects the need for white collar workers and technicians, salesmen and managers, administrative and coordinating personnel, and the scientists, engineers, and other professionals who generate and implement the knowledge and technology so important to a modern society. The increased width of the middle of the structure is amply illustrated by the fact that in 1970, 48 percent of the employed labor force was engaged in white-collar work.[15]

There are some indications, however, that the very categories of "white collar" and "blue collar" may be rendered obsolete. For the as yet small percentage of the labor force working in automated factories and offices, the historical differences between blue- and white-collar workers may be breaking down, not only because the work they do is becoming less differentiated, but also because the demands of the technology may lead employers to change their policies. Technological change has resulted in "a narrowing of traditional differentiation in terms of job content. What have hitherto been manual jobs, albeit with a high degree of skill, have an increasing conceptual content, and an increasing emphasis upon formal knowledge. On the other hand, some clerical jobs have an increasing manual content with the advent of computers." [16] Moreover, "where the ratio of capital to labour costs is high, maximum utilisation of plant becomes extremely important. It becomes even more necessary to avoid breakdowns and to have a reliable labour force. Granting such 'white-collar' conditions as an annual salary and improved fringe benefits is then seen partly as the price to pay for dependability." [17] Nevertheless, social factors rooted in the historical differences between the two groups continue to inhibit full convergence.

In some respects, the introduction of computers has served to enhance and quicken the pace of trends already occurring: the general upgrading of the skills of the labor force, the need for workers to receive more training, and the movement toward greater employment in the service sector. As computers replace men in performing more routine and standardized operations, and as they allow for more mechanized production of goods, a greater proportion of the labor force moves out of manufacturing and into the service sector, and more men need to be trained to do the jobs that cannot be performed by computers, including those associated with programming, maintaining, and operating computers.

What do these developments mean for the future of the occupational and class structure of our society? Many writers of science fiction or utopian novels have portrayed a two-class future society composed of the educated and the uneducated. John Kenneth Galbraith has expressed a similar notion: "When capital was the key to economic success, social conflict was between the rich and poor. . . . In recent times education has become the difference that divides." [18]

The increased importance of educational attainment for entry into higher-level positions does not, however, produce *ipso facto* a rigid two-class society. The technological changes that have altered the occupational distribution and increased the need for more educated manpower provide, in the long run, expanded opportunities for upward mobility as the bottom level of the occupational hierarchy contracts and the upper levels expand. And indeed, a recent large-scale study of occupational mobility in the United States finds no evidence of increasing rigidity in the occupational structure.[19]

Yet technological change in industry *has* resulted in the blocking of certain older paths of mobility. In industry, because of "the need for managerial personnel to have a broad educational and technological background, . . . a moat [has been established] between the workers and their foremen and all other supervisory personnel. It is increasingly rare for a working man to advance more than one step up the managerial ladder. He can become a foreman, but that is all." [20] Mobility in office work appears to be similarly blocked as "the middle step in the old promotion ladder"—positions requiring experience and seniority, but beneath the managerial level—appears to be growing smaller with the introduction of automation.[21] And among managerial and supervisory personnel in industry, "a 'gap' is forming between lower and higher levels of management." [22]

Such inability to move up within the hierarchy of an employing organization has been a source of frustration to many workers. To be sure, Horatio Alger myths notwithstanding, the numbers of workers who could move from low- to high-level positions has always been small. But because the criteria for such movement have become more formalized and less personal, mobility becomes increasingly difficult for those who lack appropriate certification and training. Moreover, since the blockage often results from exaggerated notions of the importance of formal education, there is an underutilization of existing talents. The goal of maintaining a fluid occupational structure coincides with the economic and technological requirements of an advanced industrial society. The failure to make the most effective use of existing talents often results in a lack of coordination between labor supply and demand that is responsible for unemployment and economic loss. The inadequacies of current education and training systems likewise help to produce unemployment by failing to keep pace with the changing demands for labor. For "while few employees actually lose their jobs when radical technological improvements are introduced, it is likely that jobs which otherwise would have been available for young people when they were ready to begin work will not be there. The suspiciously high unemployment rate for young people—about three times that for older age groups—suggests the validity of this hypothesis." [23]

One method of handling this problem would be to change the system of certification. Because on-the-job training can be used effectively to permit the existing work force to assume new roles and responsibilities, and because rapid technological change creates problems of knowledge obsolescence even among those who have been previously well trained, some analysts have been advocating the establishment of "new careers" that would allow a worker to move from a relatively unskilled entry position to more professional positions through sequences of combined on-the-job training and formal education. New forms of accreditation would be instituted in order to reward experience gained on the job. Pearl and Riessman, who advocate such change, have argued that currently "society insists that training take place prior to job placement. Such a system made sense . . . when only a small percentage of the population was engaged in highly skilled occupations, while most of the work force required little formal training. This condition no longer exists." [24]

The "new careers" concept thus focuses on those occupations in which on-the-job training could replace advance preparation. It rests also on the observation that societies need as much health, education, and welfare services as they can afford. There is room for expansion of existing careers in these service areas. Many more people could become qualified teachers, for example, if a process of moving up from the position of teacher's aide through a series of steps allowed them to become certified teachers. In addition, various social service "activities not currently performed by anyone, but for which there is a readily acknowledged need and which can also be satisfactorily accomplished by the unskilled worker" could be developed.[25] The attempt to design new types of careers is thus also responsive to the problem of providing meaningful work for the increasing numbers of workers who will not be able to find gainful employment in the labor force of the future. New public policies must be devised to overcome present rigidities in the labor market (arbitrary retirement and hiring practices, racial discrimination, relocation problems), to provide a flexible system of education and training (on-the-job training systems, sabbatical leaves for educational purposes, new mechanisms for occupational guidance), and to secure a better match between labor supply and demand.

The difficulties of effecting such changes illustrate one of the persistent themes in the relationship between technology and society: When technological change calls forth the need for adaptive mechanisms in the society, vested interests, organizational inertia, and/or traditional attitudes impede the development of such new mechanisms. In this particular case, both unions and industry, both governments and various publics present obstacles to needed social changes. Unions and industries jealously guard their powers and prerogatives and resist anything that might interfere with these. Hence, unions will often attempt to preserve the jobs of workers whose skills are becoming obsolete or in small demand, rather than to support programs of retraining. Industries will often resist job reclassifications that would require them to pay higher wage scales and oppose retraining schemes that do not offer advantages to them. Organizational inertia will prevent changes in customary hiring practices. And traditional attitudes that link work and income and that view maintenance of a job as an individual rather than a social responsibility will block significant alterations in social policy.

Problems of social choice are also involved in institutional changes of this sort. Consider the following estimate: In 1980, "given a $4,413 per capita GNP . . . achieved with a 37½-hour workweek, a 48-week workyear, and providing retraining for 1 per cent of the labor force, society could choose to retrain much more heavily (4.25 per cent of the labor force per year) or, alternatively, could add 1½ weeks per year in vacation. . . . Obviously, other choices could be made, involving a further reduction in the workweek, a lowering of retirement age, or an increased educational span for those entering the labor force." [26] The value issues involved in such choices include the following: "If the rate of technological change begins to exceed our ability to adjust to it, to what extent should the introduction of new production techniques be controlled? What price in individual freedom of action would we be willing to pay in order to eliminate unemployment among teen-age Negroes in urban ghettos? Although it seems obvious that we are becoming an increasingly leisure-oriented society, it is not nearly so apparent that we *should* become so. The increased national product resulting from a continuation of the present pattern of working hours plus increased productivity could be used to improve a wide variety of services, such as public education, that are not now adequately financed." [27]

Changes in the occupational and educational structures are not the only means by which the goal of maintaining or improving social mobility might be fostered. Various forms of income redistribution could also be employed to serve this end.[28] Yet such devices confront us with a value dilemma: a discrepancy between our proclaimed or "ideal" values and our active or "real" values. As Bennett Berger has observed, "that the system tolerates racism and poverty and urban blight is not evidence of its failure, but of its responsiveness to effective majorities and of the relatively low priorities of these problems in the eyes of these majorities. . . . That there is widespread poverty is probably regarded as deplorable by a majority of Americans. But *how* deplorable in the hierarchy of deplorables is the question. How far are we willing to go to reduce poverty? What values are we willing to sacrifice to the value of reducing poverty? Most Americans are apparently not willing to tax themselves sufficiently to provide minimum incomes. . . ." [29]

THE QUALITY OF WORK

While problems of securing adequate training and employment remain, renewed attention is being devoted to the question of whether technology affords men the possibility of doing meaningful work. To be sure, this is by no means a new question. At least since the days of Marx, men have been concerned about the "alienation" of the worker. What appears to be new is that the workers themselves are expressing publicly their dissatisfactions—both verbally and in acts of sabotage and absenteeism—and they are demanding "better working conditions," not only in the older sense of shorter working hours, higher pay, and better fringe benefits, but also work that is less monotonous, routine, and meaningless.[30] The value of work is being questioned as "rising expectations" come to include the nonmaterial aspects of life and as the Protestant Ethic appears to be losing some of its hold. Although the extent of the problem is hard to gauge, and although it has clearly not yet assumed the proportions of a social movement or a major national issue, it is a problem of considerable significance.

Ironically, demands for more interesting and meaningful work are arising at a time when there is some evidence that automation might bring a reduction in worker frustration. The modern automated plant is a far cry from the assembly line immortalized in Charlie Chaplin's *Modern Times.* Automation brings with it a decrease in the extreme division of labor characteristic of earlier industrial technology and gives the worker greater responsibility. Several studies have found that workers in automated plants are less dissatisfied or alienated than their counterparts in nonautomated plants.[31]

Yet it must be remembered that the assembly line has by no means disappeared. The technological change that has resulted in the relative decline of unskilled and standardized jobs has not eliminated the need for routine and dull work. "In the complex and diversified manufacturing sector of an advanced industrial society, at least three major kinds of blue-collar factory work exist at the same time: the traditional manual skill associated with craft technology; the routine low-skilled manual operations associated with machine and assembly-line technologies; and the 'non-manual' responsibility called forth by continuous-process technology." [32]

In the evolution of blue-collar work, the crafts continue to play some role, while the most important change is the shift from "skill" to "responsibility." What this implies for the well-being of the workers remains a matter of some dispute. Blauner's investigation of blue-collar workers in different industries concludes that because automated technology gives the worker more responsibility, greater control over the work process, and a better sense of the job as a whole, it serves to diminish his alienation. He summarizes the history of blue-collar work as follows:

> In the early period, dominated by craft industry, alienation is at its lowest level and the worker's freedom at a maximum. Freedom declines and the curve of alienation . . . rises sharply in the period of machine industry. . . . But with automated industry there is a countertrend, one that we can fortunately expect to become even more important in the future. The case of the continuous-process industries, particularly the chemical industry, shows that automation increases the worker's control over his work process and checks the further division of labor and growth of large factories. The result is meaningful work in a more cohesive, integrated industrial climate.[33]

A more recent study by Shepard of both blue- and white-collar workers confirms this conclusion.[34] Faunce similarly concludes, on the basis of a number of case studies, that automation brings a greater commitment to work, less alienation, and therefore less need for the bureaucratization that is so often both a response to worker alienation and a cause of its further growth.[35]

Wilensky takes a different position. His research on worker satisfaction has shown that jobs involving low degrees of freedom and high amounts of pressure are most conducive to alienation. He argues that automation is likely to increase rather than decrease alienation, since it brings less opportunity for interaction with co-workers, more complete predetermination of tools and techniques, more rigidly timed machine pacing, and a greater need to pay close attention to the work.[36] Some support for Wilensky's position can be found in Chadwick-Jones' study of the effects on workers of the conversion from manual to automated operations in the British steel industry. He finds that although there is a greater number of satisfied workers in the continuous-process plant than in the previous work situation, there is considerable dissatisfaction arising from the boredom and monotony of the work, the need to do shift work, and the decline in social relations on the job.[37]

Computerization does, however, bring enhanced prestige, a freedom from exhausting manual labor, and better wages to blue-collar workers. In both blue-collar and white-collar jobs, it also brings the need for greater responsibility and often for shift work. For many workers, the psychological benefits of greater responsibility are offset by the tensions it generates. The more serious consequences of error and the need for greater concentration and alertness in handling automated machinery cause considerable strain. Among white-collar workers, the need to do shift work and the requirement of greater accuracy, and hence more supervision, in their work are often significant sources of tension. Reduced salaries for clerks doing routine work, the routineness of their work, the necessity for working shifts, and the greater ability of an automated system to measure output in ways that have heretofore been limited to production workers may produce both widespread unrest and unionization among white-collar workers.[38]

Worker dissatisfaction—in automated and nonautomated industries—has long been a source of concern among social reformers. Because it impinges on the quality and quantity of work performed, it is also a source of concern to managers. The various proposals to solve or alleviate the problem of worker dissatisfaction can be seen as falling into three types: develop more creative and challenging leisure-time activities to compensate for dull jobs; provide better compensation for workers in alienating work situations; or "redesign the technology and workplace to invest work with more meaning. . . . The first solution is unrealistic; labor that requires little investment of self tends to go together with leisure that is full of malaise." The second solution is the trade-union strategy. It too is unsatisfactory because it fails to deal with the roots of the problem. The third solution involves job enlargement and job rotation. "Despite some evidence that job enlargement or rotation can both reduce job discontent and increase efficiency . . . these programs are rarely adopted in American industry. . . . To fit machine systems to the man is an idea foreign to most employers." [39]

Whether or not there is significant change in the workplace to improve the quality of work, some observers believe that the salience of work is decreasing. Whereas many critics of a decade ago envisaged a "cybernetic" world in which most traditional forms of work would be rendered obsolete, today there are images of a future in which the problem will be unwillingness to work rather than unavailability of work. Predictions about the demise of the Protestant Ethic, which were prevalent in the early 1950's, have reappeared with a vengeance.

It is asserted that increasing affluence has removed the prop of necessity, which has long supported the work ethic. As noted earlier, critics argue that the demands placed on workers to adhere to a strict Protestant Ethic come into conflict with the demands on consumers to be hedonistic.[40] Superimposed on "the worker-consumer contradiction," Reich finds another problem: "American society no longer has any viable concept of work. . . . We are no longer expected to find work happy or satisfying. There is, for example, no advertising designed to create pride in craftsmanship or in a worker's self-discipline. Nor is anyone convinced that he should work for the good of the community. Instead, the belief is created that one works only for money and status. This puts a heavy burden on money and status, a burden they are no longer able to carry." [41] Similarly, it has been contended that "the Protestant Ethic and the Puritan Temper, as social facts, were eroded long ago, and they linger on as pale ideologies, used more by moralists to admonish and by sociologists to mythologize than as behavioral realities." [42]

Kahn and Wiener sketch a variety of attitudes toward work that might exist in the year 2000, from minimal attachment to work—work as "interruption"—to maximal attachment—work as "mission." They suggest that Americans will increasingly shift toward the "interference" end of the continuum. By the year 2000, they argue, "the man whose missionary zeal for work takes priority over all other values will be looked on as an unfortunate, perhaps even a harmful and destructive neurotic. Even those who find in work a 'vocation' are likely to be thought of as selfish, excessively narrow, or compulsive." [43]

Even though the Protestant Ethic does appear to have decreased its hold on the American population, it is by no means dead. While fun-loving, novelty-seeking, sensation-oriented patterns of behavior are not hard to find, they are confined largely to the nonwork sphere and are not all-pervasive even there. Moreover, the evidence would seem to indicate that much of the cultural elite has experienced a loss rather than a gain in leisure time during the course of this century.[44] Changes in the direction of a decreased attachment to work are likely to be less rapid than many commentators assume. If one considers "all hours worked —moonlighting and the main job, overtime and straight time, as well as the increased labor force participation of women— modern economies command an impressive amount of disciplined work, and there is no sign that the trained population is suffering a markedly decreased willingness to log those hours. Moreover, the primordial meaning and function of work is dramatized by the narrow range of social contact of men squeezed out of the labor market; the aged, the school dropouts, the unemployed, and the underemployed are isolated from the mainstream of community life. Employment remains symbolic of a place among the living." [45]

PROBLEMS OF LEISURE

No such symbolism attaches to nonwork behavior. During the days of the "automation hysteria," there was much concern about how people would spend the large amounts of time that were to be freed from work obligations. As concern about massive amounts of leisure time has come to appear premature, the perception of leisure as a "problem" seems to have abated, although as with work, there has been some movement from concern about the quantity or availability of leisure to concern about its quality.

If one looks at the history of the average workweek, one finds that there has been little change over the last thirty years. The major gains in workweek reduction occurred early in the century. It is only the growth in leisure time via paid vacations that is a post–World War II development.[46] The "average hours worked per year have been declining slowly and sporadically for a long time, with the average yearly decrease about 0.3 to 0.4 per cent." [47] Over the long term, however, some increases in leisure time are likely. Although the demand for both work and leisure may vary in time, "in general the demand of most individuals for work is a declining one, whereas that for leisure is an increasing one." [48]

The distribution of free time is likely to be uneven among different occupational groups. Sectors of the upper strata today often have less leisure time than their counterparts of an earlier day. "Even though their worklives are shorter and vacations longer, these men work many steady hours week after week. . . . Considering both occupations which necessitate long hours (proprietors, some young skilled workers and foremen) and those in which men choose to work hard (professors, lawyers), there appears to be a slowly increasing minority of the male urban labor force working 55 hours a week or more." The groups experiencing increased leisure are those who "(1) have motivation and opportunity to choose leisure over income or (2) are marginal to the economy and are therefore forced into leisure." [49]

Whether the loss of leisure among some members of the upper strata is due to the requirements of the work itself or to the preferences and values of the workers, the result has often been a certain alienation or disengagement from nonwork activities comparable to the traditional lack of concern with work activities that is manifested by many blue-collar workers. As one observer has commented, "Shall we deplore the worker's lack of interest in his occupation, or the intellectual's excessive interest in work?" [50]

Most projections assume that this trend will continue, so that while workers will increasingly have free time, the upper echelons will not. If the increases in leisure time for the non-elite become significant, some serious problems might be raised by this duality. As Donald Michael has expressed it, "How do we educate one segment of society to expect to have and use productively more free time and, at the same time, educate another segment to expect to have little or no free time and not to want it?" [51] In the past, he notes, "the masses worked and the elites had free time. Now we are faced with a social inversion. . . ." [52]

Some reasons for this "social inversion" may be suggested: In a period of affluence within a society that values equality, the upper strata often must work longer hours to offset the decreasing differential between what their income will buy and what the lower strata can purchase;[53] workers have not demanded more leisure until the level of affluence was sufficient to insure that a high level of goods could be secured for their income. There is evidence that "probably for the majority of all social categories, the desire to earn more is far more compelling than the desire to have more free time." [54] Given a certain level of affluence, then, the upper strata who are devoted to their work will continue to work long hours, while those for whom work is less satisfying will choose to have more free time. Technological advances that reduce the amount of labor time required from workers support this pattern.

The perception of leisure as a "problem" is, in large part, related to this "inversion." That is, more free time for "the masses" is viewed as problematic not because they will be unable to fill it but because it is believed that they will fill it unwisely or harmfully; leisure can become a problem if many people use it in socially harmful ways. But, as Linder suggests, leisure may also be viewed as a problem if many people "occupy themselves, if not with mischief, at least with such vacuous practices as reading comics and drinking Coca-Cola. This too is something that can lead people to talk of a leisure problem. For moral, ethical, cultural, or other reasons, they cannot accept the way in which others choose to use their time." [55] Such imposition of values aside, there is some evidence that the upper strata, who tend to enjoy their work more, find themselves having a scarcity of time for all the things they want to do, while those who suffer from unhappy work lives are often unable to spend their free time in satisfying ways. Leisure can also be a social problem when large numbers of people seeking recreation find themselves in competition for scarce public resources (such as parks, beaches, roads). Congestion affects increasing numbers of people from all strata, and attempts to alleviate it have become a public issue.

As technological change has afforded greater opportunities for leisure, it has also helped to condition the ways in which free time is allocated. If "mechanization of his labor is apt to arouse the worker's distrust and hostility, the mechanization of his leisure has his full consent, not to say downright enthusiasm," as seen in the use of cars, radios, television sets, and other such equipment.[56] To be sure, residues of more traditional uses of time continue to be with us. Indeed, "certain traditional activities of the peasant and the craftsman tend more and more to become the leisure occupations of modern society that counterbalance the mechanization and rationalization of work." [57] Examples include gardening, hunting, fishing, and camping.

It is important to remember, however, that "the increased efficiency of time made possible by technology has been more than offset by growth in the demands on time, and few Americans have time hanging heavy on their hands." [58] As de Grazia has pointed out, nonworking time and leisure time are not the same thing. Travel time to and from work, household and personal maintenance, and fulfilling the various social demands and requirements of a bureaucratic society all eat into available time, while the entrance of women into the labor force causes men to share in household chores and further reduces the leisure time of the family.[59] Moreover, Linder observes, "the time devoted to enjoying different consumption goods is as essential in the consumption process as the goods themselves. . . . [T]he fact that consumption, like work and other activities, takes time [means that] a rising level of incomes [does not lead] to everyone getting more and more 'free time' and to the relaxing of the general pace of life." [60]

What increases there are in leisure time do not remain completely free. The "portion of leisure long enjoyed tends to become ordered by habitual and institutional guides, whereas new increments of leisure remain uncommitted for a while, but not necessarily permanently. . . . Of course, technological or other changes can produce profound changes in the patterns of consumption of families and hence in their patterns of leisure use, but these in turn tend to become routinized. Because of these tendencies to the routinization of the use of time and income, neither time nor income remains as discretionary as is sometimes supposed; both get committed." [61]

The future of "free" or "leisure" time will depend upon the social and individual choices about work, education, and income that determine how much free time there is and how it gets "committed." In an advanced technological society where obsolescence of knowledge appears to be an increasing problem, some form of continuous education or training might become necessary for all strata of the population. It is possible that in the future, not only the unskilled, but many occupationally skilled people will find themselves having to learn new skills and enter new careers. "This new expectation of disruptable occupational patterns will produce in some a sense of threat—who will have the ability to make it through the next educative round? Others will see it as an opportunity to try something else and to discover more about themselves." [62] In either case, there would be major infringements on available time.

It is also conceivable that as economically useful jobs become scarce in some future "leisure society," new socially useful jobs might be created. A prototype of this kind of job might be that of "hand-holder" for the aged or the sick. Persons holding such jobs would serve to provide sympathetic company for persons who otherwise have little social contact. These jobs thus would serve the dual function of filling unwanted or excessive leisure time and of providing the kind of social services that will become possible in a highly productive and affluent society.

Before any such development might come to pass—if indeed it ever does—many changes are likely to occur in the patterns of work and leisure. But the most pressing problems of the immediate future concern the alterations that will have to be made in the educational and occupational structures in order to take care of those workers who are displaced by technological change and to assure the maximum possible mesh between the needs of the individual workers and the technological and economic needs of the society at large.

Urban Problems

It is difficult these days to discuss any aspect of city life without lapsing into the rhetoric of "urban crisis." The problems attributed to the "urban crisis" generally include all those that afflict American society: crime, poverty, health, education, race relations, pollution, and so on. As James Q. Wilson has observed, "many people who use the phrase 'urban problems' know perfectly well that the problems they have in mind are not to be found exclusively in big cities (or even in cities at all) or that the problems are in every case caused or made worse by the conditions of urban life. Why, then, do they persist in using the phrase? . . . One [reason] is that a political advantage can be gained from the intellectual confusion. A concern about 'urban problems' can be shared by Negroes anxious about civil rights, intellectuals interested in poverty (or whatever), businessmen worried about downtown retail trade, mayors threatened by high local taxes, shoppers looking for a parking place, housewives fearful of purse-snatchers, and architects seeking beauty." [63]

Discussions of "technology and the city" often suffer from a similar intellectual confusion motivated by political advantage. The literature abounds in claims and counterclaims by advocates of various technological "solutions" to the "urban problem." Of course, only some problems of our cities are technological in origin or amenable to technological solutions. Moreover, it is unlikely that "any of the commonly noted 'social pathologies' marking the contemporary city can find its causes or its cure there." [64] As modern society has become increasingly urban, "the problems of the city place generated by early industrialization are being supplanted by a new array, different in kind. With but a few remaining exceptions (the new air pollution is a notable one), the recent difficulties are not place-type problems at all. Rather they are the transitional problems of a rapidly developing society-economy-polity whose turf is the nation." [65]

Nevertheless, many urban problems are in part attributable to the form and structure of today's cities, and this form has been generated by the technological changes of the recent past. Hence, the roots of at least some urban problems can be found in a mismatch between modern technologies and modern cities, and solutions to some of the difficulties plaguing the cities would appear to lie in an alteration of this relationship.

THE TECHNOLOGICAL ROOTS

It is well known that today's densely populated cities are the combined result of agricultural technologies that pushed large numbers of workers off the farms and of production technologies that drew them into factories. Specialization of, and the consequent interdependence among, manufacturing firms further stimulated the growth of centralized cities: "With relatively primitive and expensive transportation facilities, interdependence implied geographical proximity, hence centralization. . . . Technological advances in transportation and communications, largely in response to the centralizing urban trend, served to accelerate it further. Public transportation, first in the form of horsecars and trolleys, later subways and buses, facilitated intra-city transportation between home, work place, and shopping areas. The ability to draw on a working population within a greater radius of the work place permitted further growth and centralization of both industrial and commercial operations." [66]

Suburbs built before 1920 grew up around the railroad. "These suburbs, strung along a railroad line, were discontinuous and properly spaced; and without the aid of legislation they were limited in population as well as area; for the biggest rarely held as many as ten thousand people, and under five thousand was more usual. . . . Being served by a railroad line, with station stops from three to five miles apart, there was a natural limit to the spread of any particular community. Houses had to be sited 'within easy walking distance of the railroad station,' as the advertising prospectus would point out. . . ." [67]

The development and widespread use of the automobile radically changed this pattern. It was now both possible and convenient to live at a distance from the workplace. In the late 1920's and early 1930's, changes in manufacturing technology also began to serve as a force for dispersion. Newer technologies and growing plant size began to give a competitive advantage to the plant with horizontal material flows and wide aisles. Space for this kind of plant could be found only outside the central city. Improvements in transportation that reduced the cost of long-distance hauling and the growth of broader markets further facilitated dispersion of plant sites.[68]

All these developments set the stage for the rapid suburbanization that took place after World War II, "as soon as wartime controls on homebuilding and noncritical, nonresidential construction were removed." [69] As a result, "a third of wholesaling employment in the 40 largest metropolitan areas is located in suburban rings in 1963, as compared to 1/10 at the beginning of the postwar period." [70] Service employment, retailing, manufacturing, and population also continue to shift to the suburbs.

Up to this point the pattern seems to be clear and undisputed. Projections for the future, however, do not show a similar unanimity of opinion. What is probably the most prevalent view holds that current trends will continue. Metropolitan areas will continue to grow rapidly, but most of the growth will take place in the suburbs and outlying areas. As John Meyer has put it, we will see "the development of a society that is increasingly urban," with population and places of work spread evenly throughout the urban area. "This combination implies the emergence of a few very large, multi-city conurbations containing a very substantial majority . . . of the nation's population." [71] John Kain presents a slightly modified version of this general picture. He argues that jobs are moving to the suburbs faster than people are. If this continues, he suggests, the model of the future may be a "doughnut" in which workers' homes are in the central city and employment opportunities are located around the periphery.[72]

Other projections for the future of urban development include recentralization, the decline of the city as an employment and population center, and extended low-density urban sprawl. Boris Yavitz sees indications of a return to centralization stemming from the use of computers. He argues that the increasing use of computers in business has the effect of converting many city offices into "paper factories" that provide new kinds of white-collar employment for the unskilled. Computer technology also encourages some organizations to recentralize, and research and development work is probably best carried on in cities because it generally does not require much open space. Finally, city-based plants seem best suited to newer industries like electronics, which produce small instruments and components.[73]

John Kemeny suggests that the expanded use of computers will have the reverse effect. It will allow the city to become a depository of information, "a major node in the computer-communication network." Business could be transacted through this network and "tens of millions living in surrounding small towns will have continual access to these services by means of computers, television, and video phones. But they will not have to go to the city." The city could thus revert to the simpler function of being the home for a smaller number of people.[74]

Melvin Webber notes that "our compact, physical city layouts directly mirror the more primitive technologies in use at the time these cities were built. . . . If currently anticipated technological improvements prove workable," he contends, "each of the metropolitan settlements will spread out in low-density patterns over far more extensive areas than even the most frightened future-mongers have yet predicted." [75]

Thus far, the pattern of urban development seems to be a continuation of earlier trends: Urban areas continue to absorb population increases and the suburbs are growing more rapidly than the central cities. "In 1970, 149.3 million persons, 73.5 percent of the total population, were living in cities and towns (of 2,500 or more), compared with 69.9 percent in 1960. . . . [Within the 243 Standard Metropolitan Statistical Areas], for the first time the suburban population outnumbered that in the cities: 76.3 million in the suburbs against 63.9 million within the cities. Between 1960 and 1970 the proportion of the metropolitan population living in the outer rings rose from 49.5 percent to 54.4 percent." [76]

What are the problems that have resulted from the urban development described above? The poverty of inner cities in the wake of suburbanization is the most prominent. Central cities are increasingly drained of revenues as businesses and wealthier residents move to the suburbs. They cannot compete with the suburbs in attracting new business, because the suburbs provide more and cheaper land, easier access to markets, supplies, and parking facilities, and a more attractive physical and social environment. Inadequacies of housing and transportation add to the plight of the central city. There are, finally, what might be called the "quality of life" dimensions: the high population densities, the decline of the city as a cultural and commercial center, the spread of pollution, and the disappearance of "open spaces."

Population growth threatens to exacerbate these problems. Our biggest cities find that they spend more per capita as their populations increase; that is, they begin to experience "diseconomies of scale." [77] Suburban residents find that their tax dollars yield fewer of the benefits that they came to the suburbs to find.[78] According to some estimates, the United States will have a population of 300,000,000 by the end of the century, which will require a doubling of the physical plant of all its cities.[79] The problems of educating and finding jobs for new inhabitants will then become even more difficult than they are now.

While problems of providing for larger populations are not, strictly speaking, technology-related, technological changes exacerbate the situation by making it more difficult for cities to absorb new migrants. As contrasted with past migrations, "a far more difficult setting faces the migrant to the cities today." Technological progress has produced an "ever finer division of labor, calling for ever higher levels of education and training, . . . [a] shift from extractive and manufacturing industries to service industries that require long periods of preparation, . . . [and an] increasingly complex organization of the economy and polity. . . . Specialization, interdependence, and integration are the definitive traits of today's urbanism. This new scale of complexity distinguishes modern urbanism from earlier forms and is setting the policy agenda that the nation must now address. Although it is still easy to migrate to the cities, the demands of large-scale society are making it more difficult for newcomers to gain entry into the new urban society." [80]

Attempts to cope with these problems are hindered, and sometimes paralyzed, by the conflicting interests involved. The innumerable trade-offs that have to be made—between benefits to the rich and benefits to the poor, between local and national interests, between technological rationality and citizen participation—militate against any easy calculus of solution. Goals remain unclear and government policies are often contradictory. For example, urban renewal programs aimed at reviving central cities compete with home-finance provisions that make it easier to move to the suburbs. "Do we seek to raise standards of living, maximize housing choices, revitalize the commercial centers of our cities, and suburban sprawl, eliminate discrimination, reduce traffic congestion, improve the quality of urban design, check crime and delinquency, strengthen the effectiveness of local planning, increase citizen participation in local government? All these objectives sound attractive—in part, because they are rather vague—but unfortunately they are in many cases incompatible. . . . A 'revitalized' downtown business district not only implies, *it requires* traffic congestion—an 'uncongested' Broadway or State Street would be no Broadway or State Street at all. Effective local planning requires *less*, not more, citizen participation—the more views represented, the less the possibility of agreement on any single (especially any single comprehensive) view."[81]

The problem of conflicting goals and interests is rendered all the more difficult by the ineffectiveness of most local governments. The technological changes that have broadened the area of problem definitions and solutions have not yet been reflected in a broader span of governmental control. The coordination required for transportation planning, pollution control, or local economic development is not easily attained when it must be arranged through the joint efforts of numerous agencies and municipalities. Nor is innovation in either social organization or technology likely to flourish in such a situation. The determination of numerous small fiefdoms to retain their power militates against it.

SOME PROPOSED SOLUTIONS

How is a better match to be made between the state of our technology and the state of our politics? One school of thought begins with the technology and argues that there has been a failure to exploit the potential of technology for urban improvement. The failure is primarily due to the fragmentation of power that prevents the formation of a market. Urban technologies will not be developed as long as the companies involved cannot identify their customers and see an available market. Hence, there is a need for new organizations that would aggregate the requirements of a large group of customers into one market. Eberhard and others cite the School Construction System Development project as an example. This project, which involved the building of twenty-two schools in thirteen California school districts, is seen as having resulted in the innovative and beneficial use of construction technology.[82] A variant of this model, as set forth by Haire, would establish an Urban Technology Development Corporation, which would engage in research and development and provide "the neutral middle ground between users' needs and the private sector's technological capabilities." [83]

The analogy here (whether explicit or implicit) is to aerospace. Discover what the technological needs are and then mobilize the capabilities of private industry to supply them. There are few if any advocates of this scheme any more who fail to see that an urban technology mission would be far more complex than the space mission. The analogy is less to the space mission itself, of course, than to the generation of private-sector capacity to provide public goods as successfully as it has been producing goods for private consumption or for military and aerospace purposes. To achieve anything like this success in the urban area would, of course, require a good deal more money than is currently available to the cities, as well as reorganization of both industry and government.

While proponents of such schemes argue that the efficiency of the private sector would thereby be put to use for social benefit, opponents claim that the profit motive would mobilize skills and produce goods that would not be socially beneficial. Lewis Mumford, for example, has argued that the federal government should not underwrite large-scale projects of urban renewal that would attract private industry. With a large sum of money "as bait," he suggests, "a new kind of Aerospace Industry would move in, with all its typical paraphernalia of scientific research and engineering design. . . . Once started, such a scientifically ordered housing industry, commanding virtually unlimited capital . . . and providing . . . indecently large salaries and exorbitant profits, would be geared to further expansion. And it would achieve this expansion, not only by designing units prefabricated for early obsolescence, but likewise by wiping out . . . those parts of the rural or urban environment that were built on a more human plan." [84]

It is precisely the institution of a "more human plan" that advocates of private industry are claiming. The establishment of planned new communities or new towns by large corporations has been seen as one solution to the problems of urban sprawl and housing shortages. According to the Advisory Commission on Intergovernmental Relations, a "new community" is "a large scale development constructed under single or unified management, following a fairly precise inclusive plan, and including different types of housing, commercial, and cultural facilities, and amenities sufficient to serve the residents. . . ." [85] Although many builders lay claim to the designation, the commission estimates that there are no more than fifty such new communities in the entire United States.

The advocates of new communities see them as a way of draining some population from existing high-density areas and of providing their residents with a "total environment" catering to all their needs. New communities are also seen as a way of generating needed urban technologies. As the commission expresses it, "It has been suggested that a combination of three major elements are needed to produce the advanced city-building technology of which the nation . . . is capable: theoretical and applied research must be undertaken, constraints . . . on the application of new discoveries must be . . . eased, and a market for the new products and techniques must be created. New communities provide these opportunities." [86] To take advantage of these opportunities, government assistance is needed in financing and in acquiring large tracts of land. Government must also take some responsibility for the planning of these communities and for setting standards.

It has been argued that the costs of building the new communities should not be calculated without also looking at the costs of "the alternative. If we simply let our present cities double in size . . . the costs will be infinitely greater," both because of "diseconomies of scale" and because of the "social costs of urban elephantiasis." [87] Eichler and Kaplan,[88] on the other hand, see government investment in new towns as being wasteful. The money and energy should be committed instead to helping the urban poor. They reach this conclusion after examination of the efforts of a number of builders to develop such communities. Planned communities, they contend, do not deliver the social dividends they promise. They do not cure suburban sprawl; they are not more efficient; they do not shorten the journey to work; and they are not compact enough to result in savings on utilities, schools, or transportation. The opportunities that they provide for local employment and recreation are not much different from those appearing in suburbia, even though they may come at an earlier stage; and architecture and design tend to be more conservative than in unplanned developments. The most serious problem is that communities of this sort do nothing to relieve the housing problems of those with low incomes. The buyers in new communities do not want low-income families any more than the average suburban resident does.

There is also some evidence that the "urban elephantiasis" that new towns are supposed to alleviate or prevent may not be occurring. It has been pointed out that the projected increase of 100 million in the population of the United States by the year 2000 could be absorbed by the smaller cities. "If each of our smallest 200 metropolitan areas took in a half million persons, we could hold the 100 million without any of these areas exceeding 2½ million." [89] Moreover, the 1970 census reveals that "most of the nation's oldest and largest cities . . . stopped growing or showed a decline. Of the 25 largest cities, 12 lost population during the decade. . . . SMSA's [Standard Metropolitan Statistical Areas] with a population in the range of one to three million grew by 20 percent in the 1960's, whereas in those above three million the growth rate was only 10.7 percent. This seems to suggest that three million inhabitants may represent an upper limit of reasonable size, beyond which urban problems begin to become unmanageable." [90]

The advocates of new towns hold out the hope that because they give free reign to technological innovation and the entrepreneurial skills of private enterprise, they will provide answers to our urban problems and appropriate models for the cities of the future. But critics who have examined such attempts and found them wanting are led to the conclusion that the most fruitful way of dealing with our urban problems is to change our laws and institutions. Examples include the creation of regional authorities to deal with problems like transportation and pollution, and plans to share federal tax revenues with the states. The Advisory Commission on Intergovernmental Relations, the Committee on Economic Development, and others have issued reports that recommend changes in urban government for what the commission has called "civilizing the jungle of local government." [91] Recommendations include such things as "liberalizing municipal annexation procedures, . . . facilitating county consolidation, . . . granting authority for intergovernmental contracts and joint service arrangements, . . . providing for metropolitan functional authorities that offer services requiring area-wide handling, and authorizing regional councils of elected officials, . . . [and] halting the proliferation of special districts and small nonviolable units of local government in metropolitan areas." [92] Such proposals are representative of the school of thought that looks for a solution to the mismatch between urban government and technology by starting with the government.

Those who would use technology—especially in the form of advanced management techniques—to modernize and raise the efficiency of local government are increasingly coming to realize that, in the absence of such organizational changes, systems analysis or planning-programming-budgeting or computerized information systems will not accomplish very much. The movement to introduce new techniques and technologies of management into city government may, of course, become a force for consolidation of authority, although what little consolidation has occurred to date has been of information rather than of authority. The difficulty in achieving both consolidation of authority and meaningful use of new technology in local government rests in part on the uncooperativeness of those agencies and political actors who lose power as a result.

Although the initial wave of enthusiasm for the new techniques is being supplanted by a more cautious attitude and a greater questioning about their feasibility, attempts at grand-scale analysis continue. Jay Forrester's computer simulation of urban problems is an example.[93] Because cities are "complex systems," he argues, their functioning is "counter-intuitive." Programs that seem commonsensical are most likely to work *against* the solution of the city's problems. Thus, creating jobs for the unemployed, setting up training programs to upgrade skills, bringing state or federal government money into the city, and building low-cost housing will not work. Such programs fail because they attract underemployed people at rates that exceed the capacity of cities to provide either jobs, housing, or adequate public services for them. The only policy that will really promote urban revival, according to Forrester, is one in which slum housing is demolished and replaced by new business enterprise, and new enterprises are encouraged by appropriate tax policies. Only in this way can jobs be generated at a rate that exceeds the tendency of new underemployed people to flock to the city. Housing will be quite tight, but this is a price worth paying.

Several critical reviews of Forrester's work take issue with some of the underlying assumptions of the model and point out that its failure to include the suburbs as part of the system is a crucial defect.[94] John Kain argues: "If the central city reduced its low-income population by 100,000, the low-income population of the suburbs would have to increase by roughly the same amount. Although Forrester's model reflects no awareness of this aspect of metropolitan interdependence, suburban governments are all too aware of it. Indeed, much of the urban problem today is a result of suburban government's successfully pursuing precisely the kind of beggar-thy-neighbor policies Forrester advocates for the central city. . . . The solution is not, as Forrester indicates, the pursuance of narrow self-interest by each local government. Instead we need to develop a more appropriate division of responsibilities and functions among governments, and thereby remove the fiscal incentives for local governments to follow policies that, while perhaps efficient from the viewpont of narrow self-interest, are inefficient from the viewpoint of society as a whole." [95]

Failure to consider such larger social consequences is a problem also in housing and transportation. Ways need to be devised to take account of the external costs and benefits of such technologies. It has been noted that "the present cost-benefit tests of alternative transportation designs are typically based solely on estimates of direct user-benefits and of capital costs, both of which are internal to the transportation system. But the dominant consequences of a transportation system are the external ones. . . . The rub comes when we try to identify the nature of outputs from a public transportation enterprise that has no profit-and-loss accounts to guide it. Because most of the important effects are expressed only indirectly—as improved efficiency for individual firms, or as expanded opportunities for individuals, or as greater productivity in the national economy—it is difficult, perhaps impossible, to trace the many specific outputs of any given transportation investment." [96] For such reasons, the highway and private automobile have been the principal foci of our transportation policies, and such "external costs" as the inability of carless residents of the central city to get to jobs in the suburbs have been largely ignored.

The need for a holistic or "systems" approach is widely recognized, but the difficulties are also appreciated. Forrester's optimism about the computer simulation method is not widely shared. He contends that the potential for solving complex problems lies only in the development and application of such models and simulations, that these techniques will provide solutions where the more traditional methods have failed. The less optimistic and more prevalent view has been well expressed as follows: "We have . . . come far enough to appreciate what we do not know, and very few students of urban affairs would subscribe to the implicit assumption of a decade ago that if only we had the power and money to act (e.g., metropolitan-area-wide government and Federal grants) we would know what to do and how to do it. The current state of our urban transportation and urban renewal programs should force some degree of intellectual humility." [97]

Thus, despite calls for a more coherent national urbanization policy, experimentation with various alternatives in transportation, housing, and urban renewal are likely to proceed on a local basis. Policy recommendations are highly diverse. Should we allow the suburbanization movement to take its course, or should we attempt to reverse it? Should we continue to experiment with urban renewal in the attempt to revitalize the central cities, or does this only reinforce ghetto conditions? Should we instead attempt to build low-income housing outside the ghetto? [98] The answers to such questions will evolve only slowly, as efforts continue to develop a better match between modern cities and modern technologies.

□ □ □

1. *Washington Science Trends,* 23 (February 2–9, 1970), 100.
2. Alvin Weinberg, "Prospects for Big Biology," in *Research in the Service of Man: Biomedical Knowledge, Development, and Use.* A Conference sponsored by the Subcommittee on Government Research, Committee on Government Operations, United States Senate, October 24–27, 1966 (Washington, D.C.: U.S. Government Printing Office, 1967), pp. 32–43.
3. Senator Walter Mondale, Introduction of Joint Resolution to Establish a Committee on Health Science and Society, *Congressional Record,* Volume 114, No. 19, February 18, 1968.
4. Harold M. Schmeck, Jr., "Spiraling Medical Costs Reflect Tangle of Conflicting Problems," *New York Times,* April 28, 1968.
5. *Ibid.*
6. Renée C. Fox, "A Sociological Perspective on Organ Transplantation and Hemodialysis," in *New Dimensions in Legal and Ethical Concepts for Human Research,* Annals of the New York Academy of Sciences, 169 (January 1970), 410.
7. Cited in *Science News,* 93 (March 2, 1968), 220.
8. "A definition of irreversible coma. Report of the Ad Hoc Committee of the Harvard Medical School to examine the definition of brain death," *Journal of the American Medical Association,* 205 (1968), 337–40.
9. See the *New York Times,* November 23, 1969.
10. National Commission on Technology, Automation, and Eco-

nomic Progress, *Technology and the American Economy* (Washington, D.C.: Government Printing Office, 1966), p. 9.

11. *Ibid.*, p. 109.
12. James R. Bright, "The Relationship of Increasing Automation and Skill Requirements," in National Commission on Technology, Automation, and Economic Progress, *The Employment Impact of Technological Change*, Appendix Volume II to *Technology and the American Economy, op. cit.*, p. 221.
13. William A. Faunce, *Problems of an Industrial Society* (New York: McGraw Hill Book Company, 1968), p. 53.
14. *Ibid.*, p. 54.
15. U.S. Bureau of the Census, *Statistical Abstract of the United States: 1970* (Washington, D.C., 1970), p. 225.
16. Dorothy Wedderburn, "Are White-Collar and Blue-Collar Jobs Converging?" (Oberhausen, Germany, 1968), Document P 12–68 of the Third International Conference on Rationalization, Automation and Technological Change, sponsored by the Metalworkers Industrial Union of the Federal Republic of Germany, p. 11.
17. *Ibid.*, p. 20.
18. John Kenneth Galbraith, *The New Industrial State* (Boston: Houghton Mifflin Company, 1967), p. 244.
19. See Peter M. Blau and Otis Dudley Duncan, *The American Occupational Structure* (New York: John Wiley & Sons, Inc., 1967).
20. Eli Ginzberg and Hyman Berman, *The American Worker in the Twentieth Century* (New York: The Free Press, 1963), pp. 333–34.
21. Albert A. Blum, "White Collar Workers," in Irene Taviss, ed., *The Computer Impact* (Englewood Cliffs, N.J.: Prentice-Hall, Inc., 1970), pp. 69–80.
22. Elmer J. Burack and Thomas J. McNichols, "Management and Automation Research Project, Final Report" (Chicago: Illinois Institute of Technology, 1968), processed, p. 2.

23. Ginzberg and Berman, *The American Worker*, p. 353.
24. Arthur Pearl and Frank Riessman, *New Careers for the Poor* (New York: The Free Press, 1965), p. 3.
25. *Ibid.*, p. 13.
26. Juanita M. Kreps and Joseph J. Spengler, "The Leisure Component of Economic Growth," in *The Employment Impact of Technological Change*, p. 365.
27. Faunce, *Problems of an Industrial Society*, pp. 82–83.
28. See S. M. Miller and Pamela Roby, "Strategies for Social Mobility: A Policy Framework," *American Sociologist*, 6 (June 1971), 18–22.
29. Bennett W. Berger, "Strategies for Radical Social Change: A Symposium," *Social Policy*, 1 (November/December 1970), 18.
30. See, for example, Judson Gooding, "Blue-Collar Blues on the Assembly Line," *Fortune*, 82 (July 1970), 69ff.
31. See Robert Blauner, *Alienation and Freedom: The Factory Worker and His Industry* (Chicago: University of Chicago Press, 1964); Jon M. Shepard, *Automation and Alienation: A Study of Office and Factory Workers* (Cambridge, Mass.: The M.I.T. Press, 1971); and Faunce, *op. cit.*
32. Blauner, *op. cit.*, p. 169.
33. *Ibid.*, p. 182.
34. Shepard, *op. cit.*
35. Faunce, *op. cit.*
36. Harold L. Wilensky, "Work as a Social Problem," in Howard S. Becker, ed., *Social Problems: A Modern Approach* (New York: John Wiley & Sons, Inc., 1966), pp. 117–66.
37. John Chadwick-Jones, *Automation and Behaviour* (New York: John Wiley & Sons, Inc., 1969).
38. Blum, "White Collar Workers."
39. Wilensky, *op. cit.*, pp. 161–62.
40. See Charles A. Reich, *The Greening of America* (New York: Random House, Inc., 1970), and Daniel Bell, "The Cultural

Contradictions of Capitalism," *The Public Interest*, 21 (Fall 1970), 16–43.

41. Reich, *op. cit.*, p. 195.
42. Bell, *op. cit.*, p. 37.
43. Herman Kahn and Anthony J. Wiener, *The Year 2000* (New York: The Macmillan Company, 1967), p. 209.
44. See Harold L. Wilensky, "The Uneven Distribution of Leisure Time," *Social Problems*, 9 (Summer 1961), 32–56.
45. Wilensky, "Work as a Social Problem," *op. cit.*, p. 164.
46. Kreps and Spengler, "The Leisure Component," p. 365.
47. National Commission on Technology, Automation, and Economic Progress, *Technology and the American Economy*, p. 10.
48. Kreps and Spengler, *op. cit.*, p. 381.
49. Wilensky, "The Uneven Distribution of Leisure Time," *op. cit.*, pp. 55–56.
50. James R. Bright, "Technology, Business, and Education," in Walter J. Ong, S.J., ed., *Knowledge and the Future of Man: An International Symposium* (New York: Holt, Rinehart & Winston, Inc., 1968), p. 214.
51. Donald N. Michael, "Free Time—The New Imperative In Our Society," in William W. Brickman and Stanley Lehrer, eds., *Automation, Education, and Human Values* (New York: School and Society Books, 1966), p. 303.
52. *Ibid.*, pp. 300–301.
53. See Staffan B. Linder, *The Harried Leisure Class* (New York: Columbia University Press, 1970), pp. 34–37.
54. Joffre Dumazedier, *Toward a Society of Leisure* (New York: The Free Press, 1967), p. 42.
55. Linder, *op. cit.*, p. 12.
56. Dumazedier, *op. cit.*, p. 46.
57. *Ibid.*, pp. 59–60.
58. Kreps and Spengler, "The Leisure Component," p. 366.

59. Sebastian de Grazia, *Of Time, Work, and Leisure* (Garden City, N.Y.: Doubleday & Company, Inc., 1964).
60. Linder, *op. cit.*, p. 77.
61. Kreps and Spengler, *op. cit.*, p. 383.
62. Donald N. Michael, *The Unprepared Society* (New York: Basic Books, Inc., Publishers, 1968), p. 43.
63. James Q. Wilson, "Urban Problems in Perspective," in James Q. Wilson, ed., *The Metropolitan Enigma* (Washington, D.C.: Chamber of Commerce of the United States, 1967), p. 319.
64. Melvin M. Webber, "The Post-City Age," *Daedalus*, 97 (Fall 1968), 1093.
65. *Ibid.*, p. 1092.
66. Boris Yavitz, "Technological Change," in Eli Ginzberg, ed., *Manpower Strategy for the Metropolis* (New York: Columbia University Press, 1968), pp. 46–47.
67. Lewis Mumford, *The City in History* (New York: Harcourt Brace Jovanovich, Inc., 1961), p. 504.
68. A more extended account of these developments can be found in Yavitz, "Technological Change," *loc. cit.*
69. John Kain, "The Distribution and Movement of Jobs and Industry," in James Q. Wilson, ed., *The Metropolitan Enigma* (Washington, D.C.: Chamber of Commerce of the United States, 1967), p. 9.
70. *Ibid.*, p. 22.
71. John R. Meyer, "Regional and Urban Locational Choices in the Context of Economic Growth," Discussion Paper Number 9, Program on Regional and Urban Economics, Harvard University, p. 13.
72. Kain, *op. cit.*, p. 25.
73. See Yavitz, "Technological Change."
74. John G. Kemeny, "The City and the Computer Revolution," in *Governing Urban Society: New Scientific Approaches*, Monograph No. 7, American Academy of Political and Social Science, 1967, p. 62.

75. Webber, "The Post-City Age," p. 1098.
76. Philip M. Hauser, "The Census of 1970," *Scientific American*, 225 (July 1971), 20.
77. See Advisory Commission on Intergovernmental Relations, *Urban and Rural America: Policies for Future Growth* (Washington, D.C.: U.S. Government Printing Office, 1968).
78. See Statement of the National Committee on Urban Growth Policy, in Donald Canty, ed., *The New City* (New York: Praeger Publishers, Inc., for Urban America, Inc., 1969), pp. 169–74.
79. See John Fischer, "Planning for the Second America," *Harper's*, 239 (November 1969), 21–26.
80. Webber, *op. cit.*, p. 1104.
81. James Q. Wilson, "The War on Cities," *The Public Interest*, 3 (Spring 1966), 29.
82. See John P. Eberhard, "Technology for the City," *International Science and Technology*, 57 (September 1966), 18–29.
83. See Mason Haire, "Industrial Technology and Urban Affairs," *Technology Review*, 71 (February 1969), 22–27.
84. Lewis Mumford, *The Urban Prospect* (New York: Harcourt Brace Jovanovich, Inc., 1968), pp. 225–26.
85. Advisory Commission on Intergovernmental Relations, *Urban and Rural America*, p. 64.
86. *Ibid.*, p. 101.
87. Fischer, "Planning for the Second America," p. 26.
88. Edward P. Eichler and Marshall Kaplan, *The Community Builders* (Berkeley and Los Angeles: University of California Press, 1967).
89. William Alonso, "The Mirage of New Towns," *The Public Interest*, 19 (Spring 1970), 6.
90. Hauser, "The Census of 1970," p. 20.
91. See Advisory Commission on Intergovernmental Relations, *Urban America and the Federal System* (Washington, D.C., October 1969); *Metropolitan America: Challenge to Fed-*

eralism (Washington, D.C., 1966); and Committee for Economic Development, *Modernizing Local Government* (New York, 1966).

92. Advisory Commission on Intergovernmental Relations, *Urban America and the Federal System*, pp. 4–5.
93. Jay W. Forrester, *Urban Dynamics* (Cambridge, Mass.: The M.I.T. Press, 1969).
94. See John F. Kain, "A Computer Version of How a City Works," *Fortune*, 80 (November 1969), 241–42; and Herbert Weinblatt, "Urban Dynamics: A Critical Examination," *Policy Sciences*, 1 (1970), 377–83.
95. Kain, *op. cit.*, p. 242.
96. Melvin M. Webber and Shlomo Angel, "The Social Context for Transport Policy," Committee on Science and Astronautics, U.S. House of Representatives, *Science & Technology and the Cities* (Washington, D.C.: U.S. Government Printing Office, 1969), p. 59.
97. Wilbur R. Thompson, "On Urban Goals and Problems," in U.S. Congress Joint Economic Committee, *Urban America: Goals and Problems* (Washington, D.C., August 1967), p. 114.
98. See John F. Kain and Joseph J. Persky, "Alternatives to the Gilded Ghetto," *The Public Interest*, 14 (Winter 1969), 74–87.

Selected Bibliography

The following list of recent publications on technology and society contains books only. Many of the more important articles in this area appear as references in the body of the text. The books listed below by Ferkiss, Mesthene, and Baier and Rescher contain lengthy bibliographies.

GENERAL AND HISTORICAL

Bell, Daniel, ed., *Toward the Year 2000* (Boston: Beacon Press, 1969).

Boulding, Kenneth E., *The Meaning of the 20th Century* (New York: Harper & Row, 1964).

Ellul, Jacques, *The Technological Society* (New York: Alfred A. Knopf, 1967).

Faunce, William A., *Problems of an Industrial Society* (New York: McGraw-Hill Book Company, 1968).

Ferkiss, Victor C., *Technological Man: The Myth and the Reality* (New York: George Braziller, 1969).

Giedion, Siegfried, *Mechanization Takes Command* (New York: W. W. Norton & Co., 1969).

Kahn, Herman, and Wiener, Anthony J., *The Year 2000* (New York: The Macmillan Company, 1967).

Mesthene, Emmanuel G., *Technological Change: Its Impact on Man and Society* (Cambridge, Mass.: Harvard University Press, 1970; New York: Mentor Books, 1970).

Morison, Elting E., *Men, Machines, and Modern Times* (Cambridge, Mass.: The M.I.T. Press, 1966).

Mumford, Lewis, *The Myth of the Machine*. Volume I: *Technics and Human Development* (New York: Harcourt Brace Jovanovich, 1967). Volume II: *The Pentagon of Power* (New York: Harcourt Brace Jovanovich, 1970).

Schon, Donald A., *Technology and Change* (New York: Dell Publishing Co., 1967).

———, *Beyond the Stable State* (New York: Random House, 1971).

White, Lynn, Jr., *Medieval Technology and Social Change* (New York: Oxford University Press, 1962).

POLITICAL AND ECONOMIC ORGANIZATION

Brooks, Harvey, *The Government of Science* (Cambridge, Mass.: The M.I.T. Press, 1968).

Drucker, Peter F., *The Age of Discontinuity* (New York: Harper & Row, 1969).

Galbraith, John Kenneth, *The New Industrial State* (Boston: Houghton Mifflin Co., 1967).

Gilpin, Robert, and Wright, Christopher, eds., *Scientists and National Policy-Making* (New York: Columbia University Press, 1964).

Harrington, Michael, *The Accidental Century* (Baltimore: Penguin Books, Inc., 1967).

Heilbroner, Robert L., *The Limits of American Capitalism* (New York: Harper & Row, 1966).

Kariel, Henry S., *The Promise of Politics* (Englewood Cliffs, N.J.: Prentice-Hall, Inc., 1966).

Lakoff, Sanford A., ed., *Knowledge and Power* (New York: The Free Press, 1966).

Mansfield, Edwin, *The Economics of Technological Change* (New York: W. W. Norton & Co., 1968).

Meynaud, Jean, *Technocracy* (London: Faber and Faber, 1968).

Michael, Donald N., *The Unprepared Society* (New York: Basic Books, Inc., 1968).

National Commission on Technology, Automation, and Economic Progress, *Technology and the American Economy* (Washington, D.C.: Government Printing Office, 1966), and the six appendix volumes thereto.

Nelson, Richard R.; Peck, Merton J.; and Kalachek, Edward D.; *Technology, Economic Growth and Public Policy* (Washington, D.C.: The Brookings Institution, 1967).

Nelson, William R., ed., *The Politics of Science* (New York: Oxford University Press, 1968).

Nieburg, H. L., *In the Name of Science* (Chicago: Quadrangle Books, 1966).

Price, Don K., *The Scientific Estate* (Cambridge, Mass.: Harvard University Press, 1965).

Reagan, Michael D., *Science and the Federal Patron* (New York: Oxford University Press, 1969).

Schmookler, Jacob, *Invention and Economic Growth* (Cambridge, Mass.: Harvard University Press, 1966).

U.S. House of Representatives Committee on Science and Astronautics, *Technology: Processes of Assessment and Choice*, Report of the National Academy of Sciences (Washington, D.C.: Government Printing Office, 1969).

CULTURE AND PSYCHOLOGY

Baier, Kurt, and Rescher, Nicholas, eds., *Values and the Future: The Impact of Technological Change on American Values* (New York: The Free Press, 1969).

Bennis, Warren G., and Slater, Philip E., *The Temporary Society* (New York: Harper & Row, 1968).

Fromm, Erich, *The Revolution of Hope: Toward a Humanized Technology* (New York: Bantam Books, 1968).

Green, Thomas F., *Work, Leisure, and the American Schools* (New York: Random House, 1968).

Hacker, Andrew, *The End of the American Era* (New York: Atheneum, 1970).

Hook, Sidney, ed., *Human Values and Economic Policy: A Symposium* (New York: New York University Press, 1967).

Keniston, Kenneth, *The Uncommitted: Alienated Youth in American Society* (New York: Harcourt Brace Jovanovich, Inc., 1965).

Marcuse, Herbert, *One-Dimensional Man* (Boston: Beacon Press, 1964).

Marx, Leo, *The Machine in the Garden: Technology and the Pastoral Ideal in America* (New York: Oxford University Press, 1964).

McLuhan, Marshall H., *Understanding Media* (New York: New American Library, 1966).

Mishan, E. J., *Technological Growth: The Price We Pay* (New York: Praeger Publishers, 1970).

Reich, Charles, *The Greening of America* (New York: Random House, 1970).

Roszak, Theodore, *The Making of a Counter Culture: Reflections on the Technocratic Society and Its Youthful Opposition* (New York: Doubleday & Company, Inc., 1969).

Sypher, Wylie, *Literature and Technology: The Alien Vision* (New York: Random House, 1968).

Toffler, Alvin, *Future Shock* (New York: Random House, 1970).

White, Lynn, Jr., *Machina Ex Deo* (Cambridge, Mass.: The M.I.T. Press, 1968).

COMPUTERS AND INFORMATION TECHNOLOGY

Bogulslaw, Robert, *The New Utopians: A Study of System Design and Social Change* (Englewood Cliffs, N.J.: Prentice-Hall, Inc., 1965).

Greenberger, Martin, ed., *Computers, Communications and the Public Interest* (Baltimore: Johns Hopkins Press, 1971).

MacBride, Robert O., *The Automated State: Computer Systems as a New Force in Society* (Philadelphia: Chilton Book Company, 1967).

Martin, James, and Norman, Adrian, *The Computerized Society* (Englewood Cliffs, N.J.: Prentice-Hall, Inc., 1970).

Pylyshyn, Zenon, ed., *Perspectives on the Computer Revolution* (Englewood Cliffs, N.J.: Prentice-Hall, Inc., 1970).

Rhee, H. A., *Office Automation in Social Perspective* (Oxford: Basil Blackwell, 1968).

Sackman, Harold, *Computers, System Science, and Evolving Society* (New York: John Wiley & Sons, 1967).

Sackman, Harold, and Nie, Norman, eds., *The Information Utility and Social Choice* (Montvale, N.J.: AFIPS Press, 1970).

Simon, Herbert A., *The Shape of Automation for Men and Management* (New York: Harper & Row, 1965).

Taviss, Irene, ed., *The Computer Impact* (Englewood Cliffs, N.J.: Prentice-Hall, Inc., 1970).

Westin, Alan F., ed., *Information Technology in a Democracy* (Cambridge, Mass.: Harvard University Press, 1971).

Whisler, Thomas L., *The Impact of Computers on Organizations* (New York: Praeger Publishers, 1970).

Wilensky, Harold L., *Organizational Intelligence* (New York: Basic Books, Inc., 1967).

Withington, Frederic, *The Real Computer: Its Influence, Uses, and Effects* (Reading, Mass.: Addison-Wesley Publishing Co., 1969).

Index